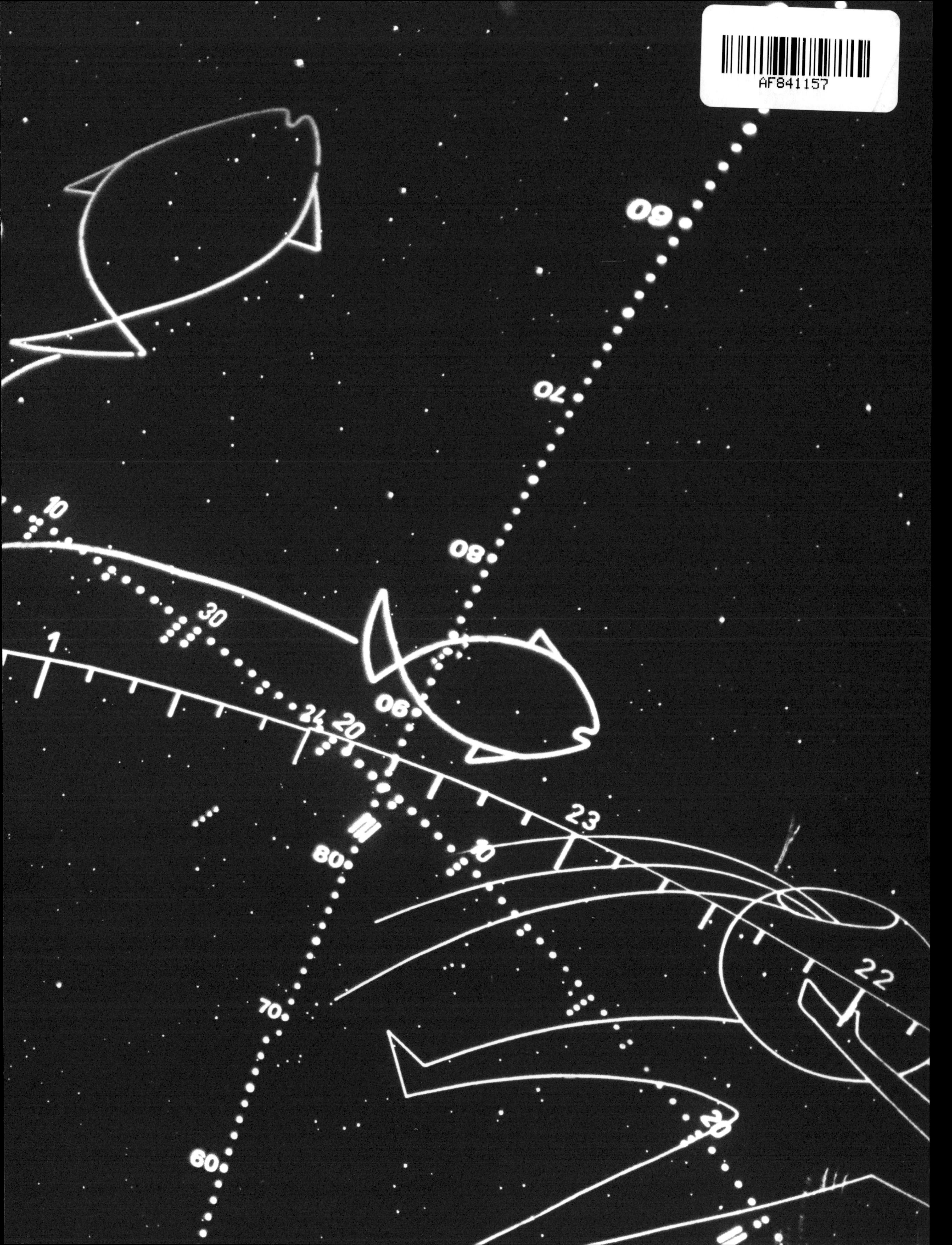

10
30
1
20
23
22
80
70
60

DIRECTION—SPACE!

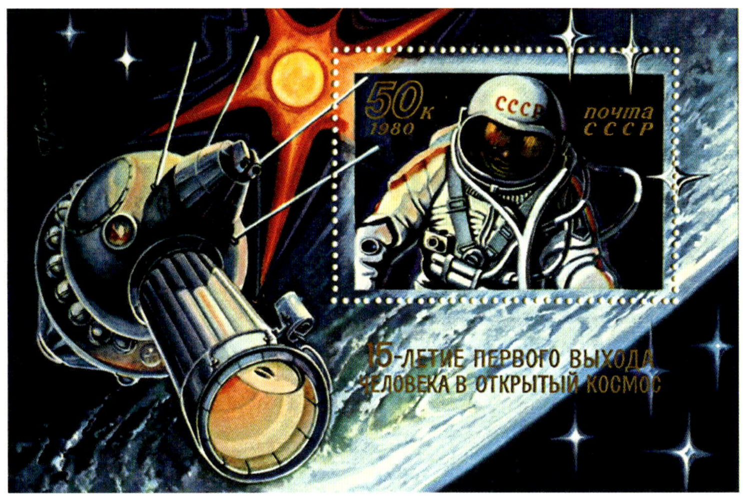

To my grandfather, a man of his word

MARIA GRUZDEVA **DIRECTION—SPACE!**

with an essay by Matthew Shaul

dewi lewis publishing

On 12 April 1961, man's first flight into space took place. Yuri Gagarin became the first man in outer space and the first to orbit the Earth.

The history of cosmonautics has a rich past: the first manned flight was followed by a number of extraordinary achievements: the launch of automated space missions to different planets, the first extra-vehicular activity, the construction of the International Space Station and the Interkosmos space exploration programme. Yet, the ultimately humanist ideals underpinning these achievments – the desire to explore and expand the boundaries of human knowledge – have inevitably been tainted with the rhetoric of politics, the space race, the arms race, and the Cold War.

Once you have entered this world, you begin to understand that it is not aggressive or inimical. Its ideals are pure. It is patriotic and international at the same time. It is the people who make it what it is: constructors, instructors, cosmonauts and their families, dynasties who pass on their skills and values from one generation to another. Children play with toy-rockets and already know what they want to be when they grow up.

These photographs were taken at the Yuri Gagarin Russian State Science Research Cosmonauts Training Centre, also known as Star City – both a military research facility and a city, highly secret in the Soviet era, and where cosmonauts have lived and trained since the 1960s – and at Baikonur, the world's first and largest space launch facility. These places hold the weight of the past with its traces of history, power and with a ghost-like presence that remains. They also continue to shape the reality of the people who live and work there now.

Generations of cosmonauts have trained in these surroundings and the physical space and its spirit have been preserved. Reticence and insularity explain the unity of this world: people exist as a tight community, a family that works towards the same aim. It is a model of sincerity and faith, something which has largely been lost in our contemporary world. Their love for Space is a form of love for their native Earth. Their aspiration and direction is into space. The air they breath is filled with star dust.

Maria Gruzdeva

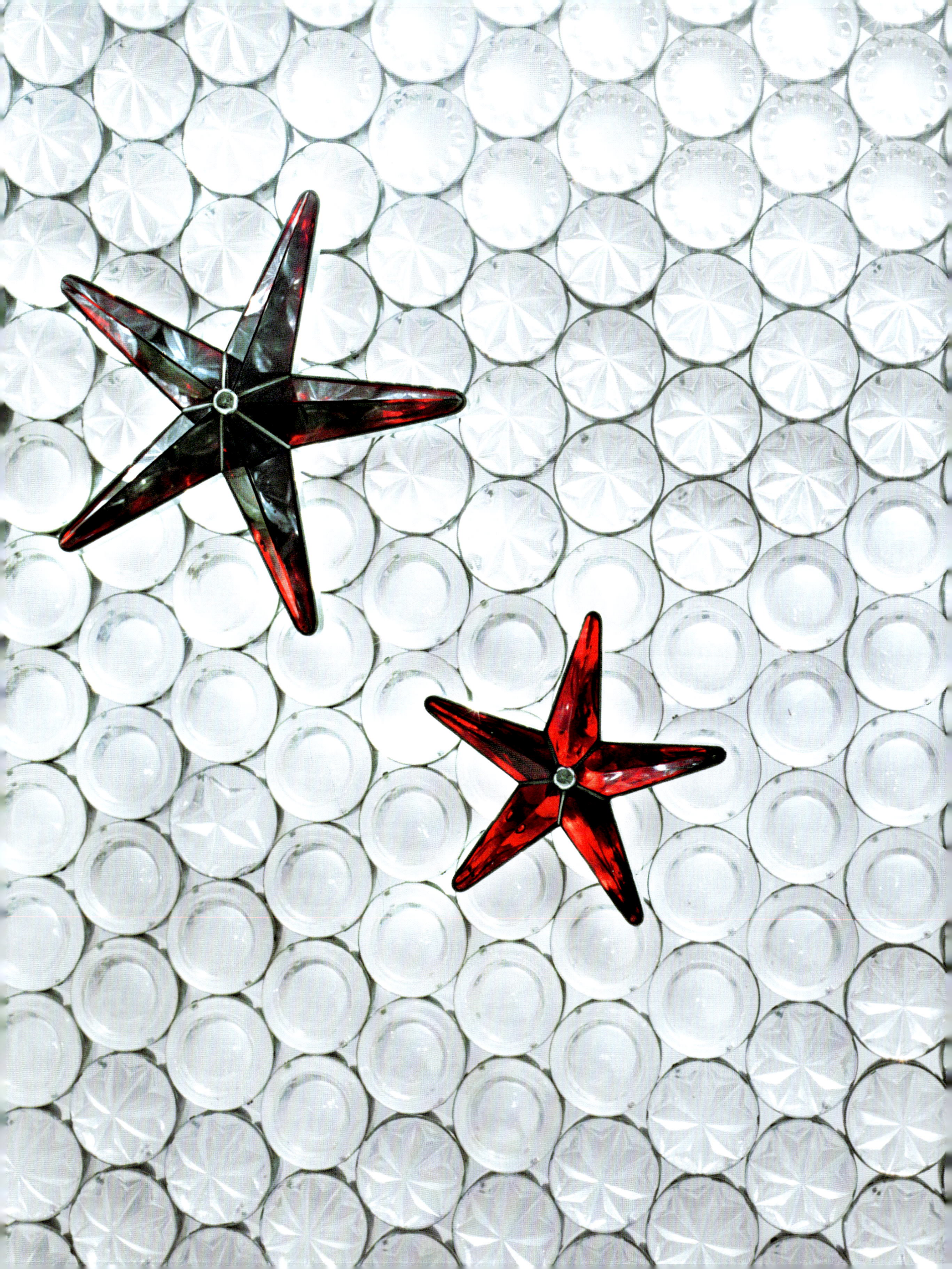

UNITY OF PURPOSE

Sputnik, Soyuz, Vostock, Lunakhod. For a kid who grew up as a self-proclaimed space fanatic the names are the stuff of legend. Viewed from the West and eclipsed by NASA's slick, shiny PR skills, the Soviet space programme was fascinating but somehow tantalisingly out of reach.

My *ABC Book of Space* (purchased ca. 1970) devoted but a few lines to Soviet space exploration, its principal interest being Valentina Tereschkova the first woman in space. In general, with the resources and picture libraries of the Western media at its disposal, the book gave the Soviet alternative to Nasa's celestial hegemony short shrift, stating only that 'foreigners were banned from the Soviet space launch facility at Baikonur', which was located somewhere amongst the unknowable steppes of Soviet Central Asia.

As a space nerd, however, I remained fascinated by certain glaring differences between the two space programmes. Firstly, why were the 'Russians' called cosmonauts when the Americans were called astronauts?; why, when they returned to earth did the Russians touch down on land when the Americans (in the presence of the massed legions of the world's media) plunged into the sea?; was the 'space race' actually a race?; was there some kind of finish line and why, if there was, did the Russians seem unerringly to keep crossing it before the Americans?

Given the jealousy with which both sides guarded the secrets of their missile development programmes, and the limitations that were placed on Soviet procurement as the dollar-driven Western economies ran riot after World War Two, the list of firsts achieved by the Soviet space programme is extraordinary and includes the first artificial satellite (*Sputnik*, 1957), the first man in space (Yuri Gagarin, 1961), the first robotic space rover (*Lunokhod 1* – the Moon, 1970) and the first data received from another planet (*Venera 7* – Venus, 1970).

Beyond these impressive milestones however, I think what always fascinated me was the style and the designs of the Soviet space programme: I distinctly remember the BBC children's news programme *Newsround* (accurately) comparing the *Lunokhod* lunar rover to a Victorian bath tub as it explored the Lunar suface in 1970. There was a 'make do and mend' characteristic, but equally something enduringly elegant in the designs of the Soyuz rocket system, with four trade-mark liquid fuel boosters clinging to its base like the roots of a tree.

Decorative mosaic, detail.

Filtered by the protective mantle of Western media and propaganda, my own knowledge of the realities of life in the Soviet Union was always viewed through my experience of, and my personal friendships in, the former German Democratic Republic, the Soviet Union's staunchest European ally and the Eastern Bloc's economic 'over-achiever'.

The cloistered world that Maria Gruzdeva has captured at Baikonur and Star City evokes the uncomplicated, practical and functional characteristics of a design culture, which was astonishingly inventive given the difficulties the Soviet Bloc encountered in securing raw material and Western micro technology payable in hard currency.

And this design culture was by no means restricted to Soviet high technology. One of the trademarks of east European interior design from Berlin to Lubljana, as well the House of Cosmonauts at Star City – and which to Western eyes is blocky and slightly cumbersome – is the decorative pressed aluminium ceiling tile. Equally the plush leather or velvet sofa, or the office furniture which looks like a throwback to the 1970s, conjure up the aesthetics of a simpler era. These have survived and are still used in the 'former East' – because a lack of replaceable resources and the absence of a throwaway culture is a great preserver.[1]

Equally, in socialist communication design, there is a thread which runs through disciplines as diverse as children's television[2], graphic and urban design, and which is shared by the emblems and memorabilia of the Soviet space programme.

Celebrated in postage stamps or produced as medals, badges and lapel pins using lightweight aluminium alloys, these ephemeral – and now collectible – PR devices have a naïve graphic purity, which draws heavily on the iconography of rocket propulsion, the vapour trail, the flags of fraternal nations and the stars, and which is possessed of a child-like wonder.

Used effectively by the Soviet space programme's fledgling PR machine this iconography was the standard-bearer for a campaign which achieved global media penetration in the years prior to Apollo 11's landing on the Moon in July 1969, with the fresh faced good looks of Yuri Gagarin – the first man in space – as its principal poster boy.

Internationalist in character, one of the main achievements of the Soviet space programme was the opportunity it extended to citizens of fraternal and 'sympathetic' countries such as Cuba, France and India to participate in the space programme. The value of this collective human enterprise, as Gruzdeva has pointed out, transcended politics.

Writing in a more uncertain, conflicted and less predictable age it seems extraordinary and rather wonderful that the Soviet postal service thought it appropriate to celebrate the collective endeavour of the successful Soyuz-Apollo test flight in 1975 with a postage stamp even when, most notably in Vietnam, the Soviet Union and the United States remained at geopolitical loggerheads.

Adopting an approach to photography nebulously referred to as 'fine art photography' Maria Gruzdeva brings profoundly personal insights to a pictorial journey, which it might lack were it a straight documentary piece commissioned for a colour magazine. Having clearly gained the trust and the consent of her portrait subjects she seems to have become an almost unnoticed observer of the narratives being played out before her. As an artist, rather than a journalist, Gruzdeva brings us the totality of Russia's space programme. With an eye for both the historical fanfare of technological achievement *and* what it means to those who deliver these achievements, she conveys profound insights for the 'human-ness' and 'lived in' character of the space programme in the private moments and personal mementoes, or the newspapers abandoned on control consoles.

With a genuine pride for what they are doing, her subjects resonate enthusiasm for their training and expertise, for the audacity of their ambitions and for the unity of their purpose, pointing to an often forgotten characteristic of Soviet (and post Soviet) social organisation – the opportunity for citizens of either sex from all walks of life to aspire to the career they wanted – in this case as the space programme's cosmonauts, technicians or physicians.

Gruzdeva has spoken of the physical isolation of Baikonur and the social isolation of Star City's (formerly top secret) training facility. Fascinated by closed communities and shared belief systems she has suggested that what drew her to the space programme was the almost 'monastic' belief in 'making the impossible possible' [3] – the idealistic spirit which clearly persists to this day. And this idealism is all the more extraordinary when we consider that, throughout most of its history, the achievements of the space programme were realised against a backdrop of profound political autocracy, ideological conceit and economic paralysis in wider Soviet society.

The public monuments at Baikonur and Star City share an aesthetic kinship with the angular sculptures, murals and roughly hewn mosaics, which are (or were) prominent facets of urban design across Eastern Europe. In the aftermath of the revolutions, which swept communists from power across the Soviet Empire between 1988 and 1991, public sculptures of Lenin, Marx and Stalin were torn down or removed by municipal decree.

Yet the space programme's monuments, idealism and unity of purpose have not only survived the demise of the Soviet Union but, at time of writing (January 2011), seem to be thriving. The Russian Space Agency (operating from the Baikonur Cosmodrome, which is now rented by the Russian Space Agency from the Republic of Kazakhstan) freely collaborates with the European Space Agency. Operating with lean economic efficiency (and as a nice little earner), it is also currently the only way for American astronauts to reach the International Space Station, following the retirement of the Space Shuttle.[4]

Similar humanist endeavours across Eastern Europe have fallen victim to sweeping political changes and new economic orthodoxies. I was reminded immediately by Gruzdeva's photographs of the widely documented idealism, which underpinned the construction of East Berlin's *Palast der Republik* (Palace of the Republic). This edifice – a 70s architectural icon – graced Berlin's Museum Island between 1976 and its closure in 1990. Although commissioned and built by the GDR's morally bankrupt Stalinist regime, it was an enterprise in whose construction workers from all over the GDR had participated and was a genuinely well-loved cultural centre and public space. Like so many monuments from Germany's short socialist history – some claim in a bid to erase that history – it was controversially demolished in the mid 2000s.[5]

Conversely, perhaps the urge to sweep away the institutions of the space programme whose values, as Gruzdeva has put it, are 'humanist and universal' was not as strong as with institutions and state agencies more instrumentally linked to communist ideology. Perhaps this is because the essence of space exploration is collegial and humanist in that it demands concerted action and because its lofty ambitions transcend politics and the nation state.

As Gruzdeva aptly puts it in her introduction – 'Once you have entered this world, you begin to understand that it is not aggressive or inimical. Its ideals are pure. It is patriotic and international at the same time.'

Matthew Shaul

Matthew Shaul is Creative Director of Departure Lounge Contemporary Art and Media, Luton, and an independent curator.

[1] Further examples of the GDR's highly individual design culture can be found in *Bild der Heimat* (Picture of Home) Erasmus Schröter (ed), Schwarzkopf & Schwarzkopf, Berlin, 2002 (ISBN 3-89602-421-3) a fascinating journey through East German interior and urban design told through its remarkable and ideologically driven holiday postcard industry.

[2] *Der Sandmann* (The Sandman) the central character in one of East Germany's most successful children's TV shows emphasised the value of collective enterprise. Travelling the world and the stars with friends from across the Soviet bloc his adventures were served up with an ideological flavour. His preferred mode of transport being aircraft and space vehicles (generously decorated with the flags of fraternal states) whose designs come straight out of a Soviet stamp collection.

[3] Telephone interview with the photographer, December 2010.

[4] BBC News, 7 October 2010.

[5] For further insights into the history of Berlin's Palace of the Republic please see Holfelder, Moritz, *Palast der Republik, Aufstieg und Auffall eines Symbolischen Gebäudes,* Ch LinksVerlag, Berlin, 2008.

Gates leading to the garage, petrol station and repair centre of Star City.

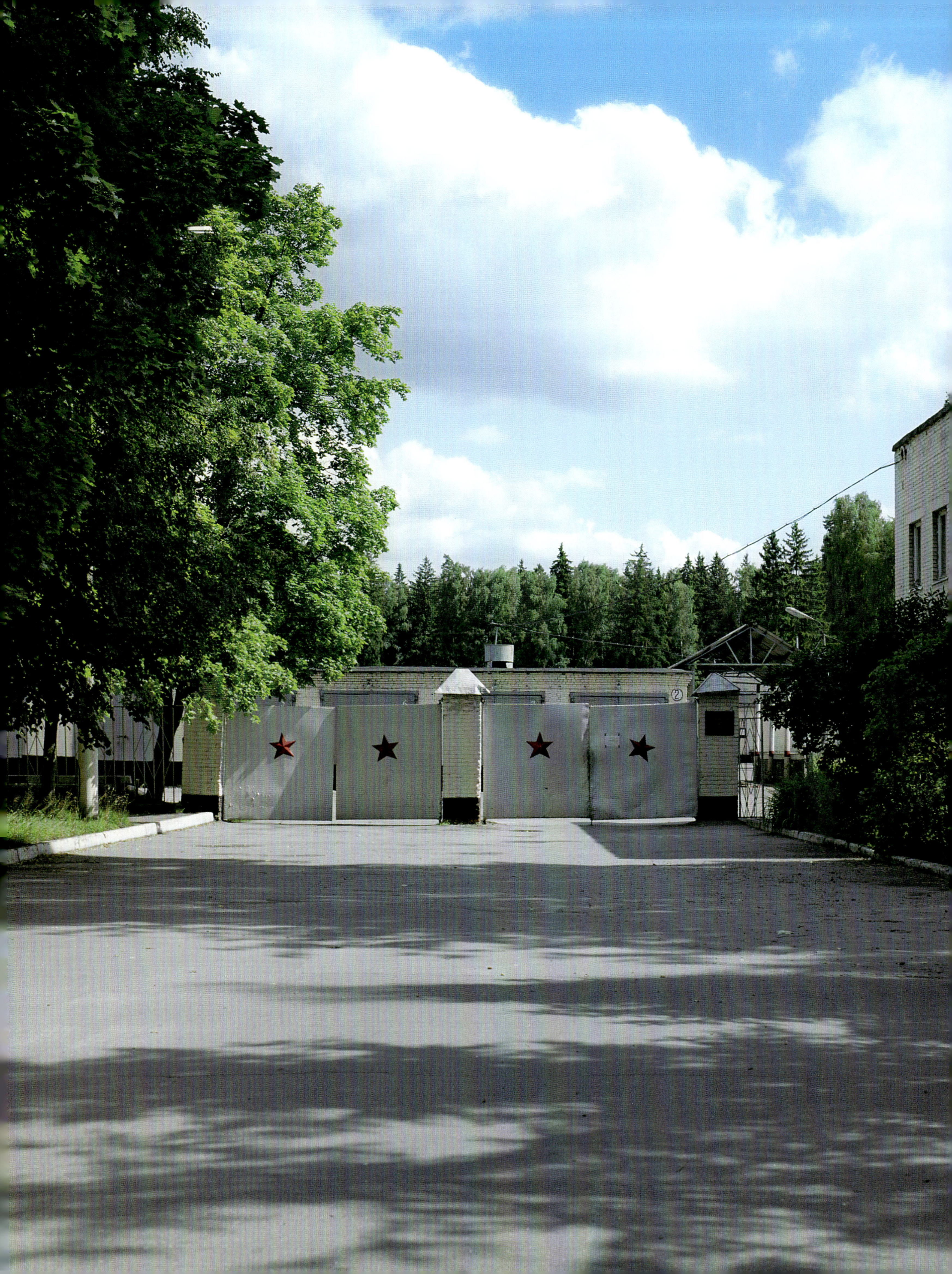

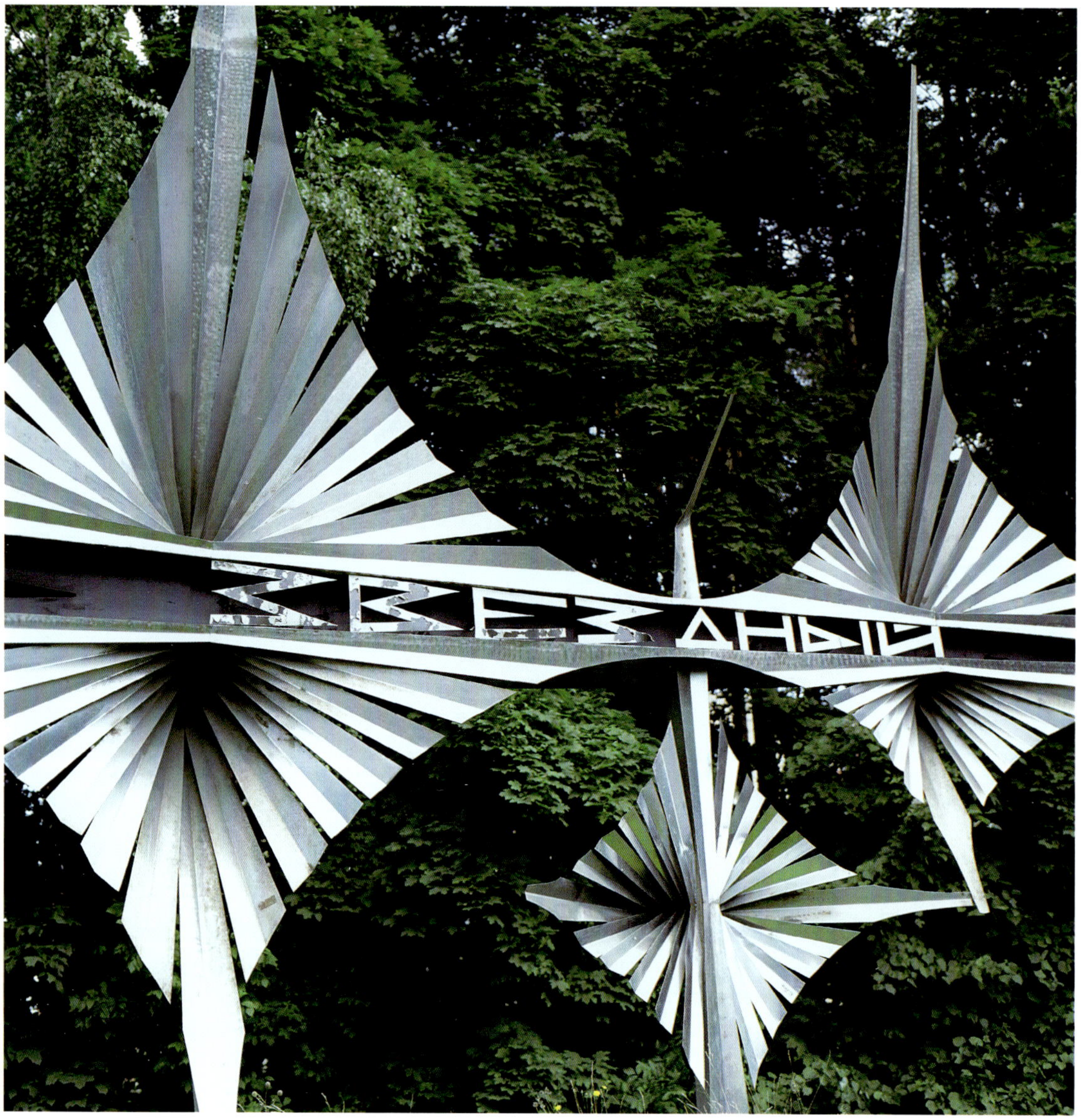

Above: Star City road sign.

Right: Main headquarters building, Star City.

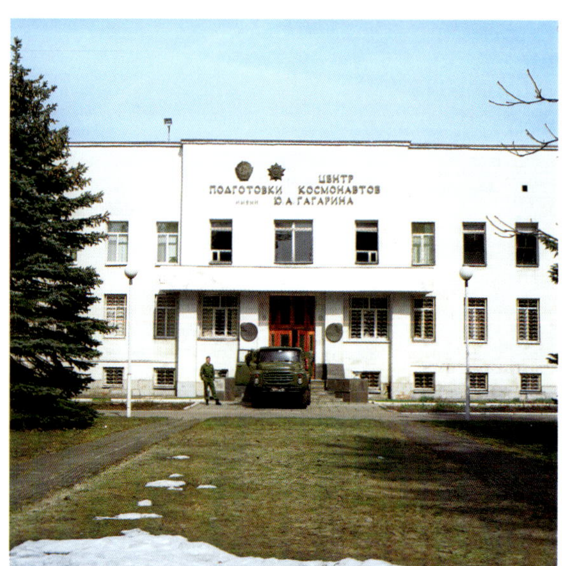

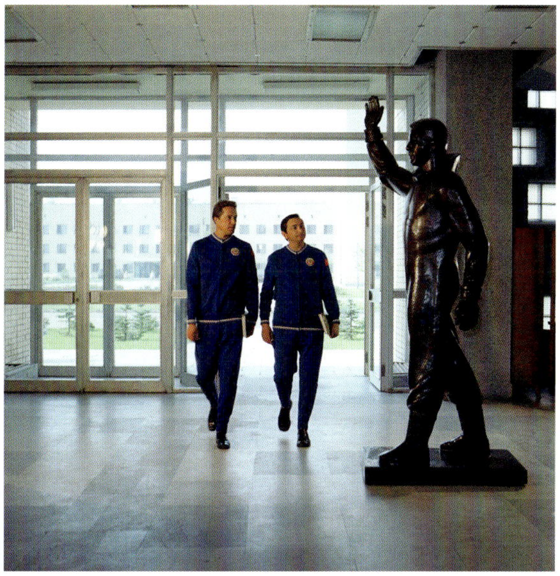

Left: Cosmonauts, V F Bykovsky and V V Aksyonov, space flight crew of Soyuz–22, launched 15 September 1976.

Below: Monument to Y A Gagarin, 2nd Department hallway.

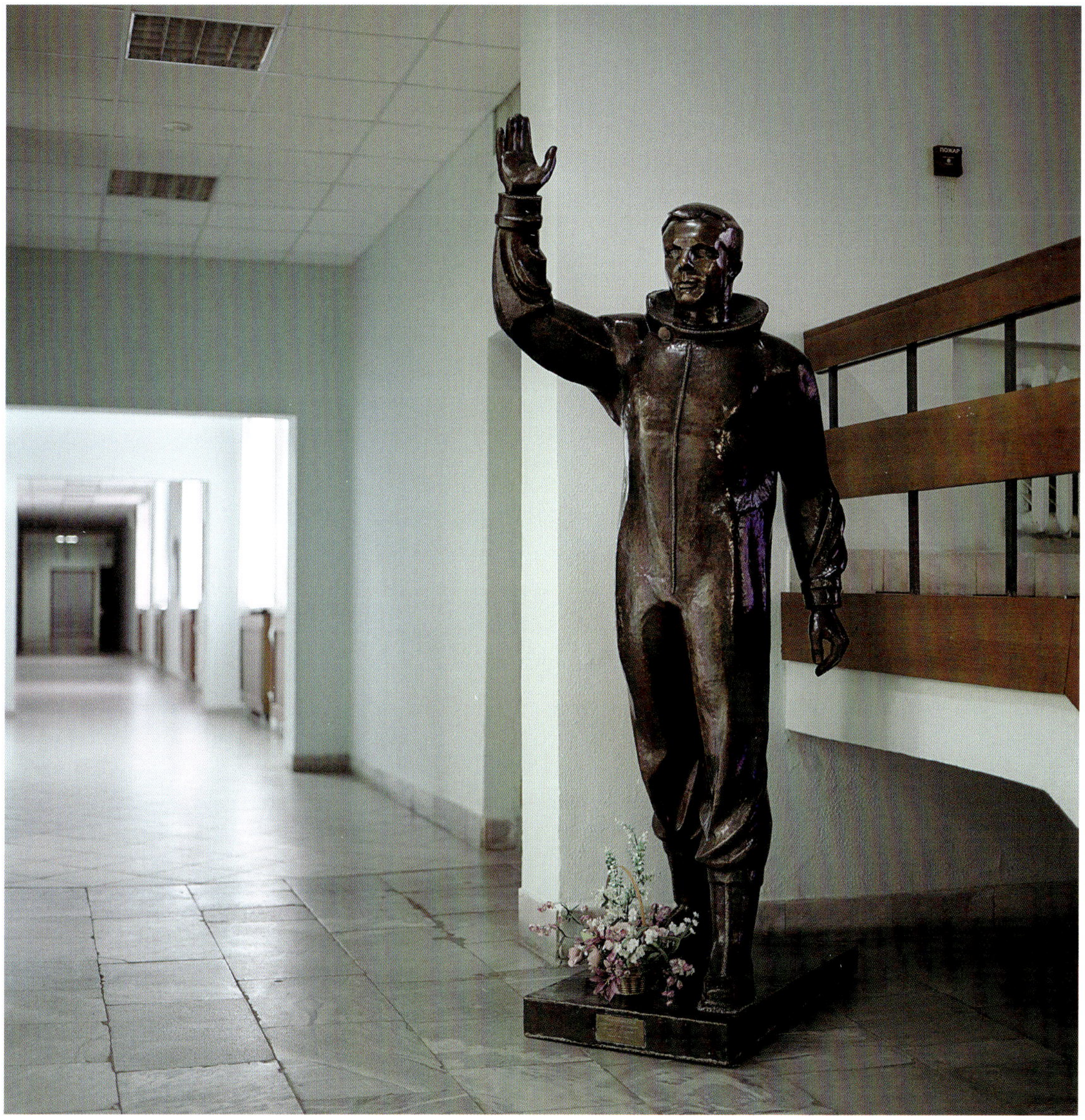

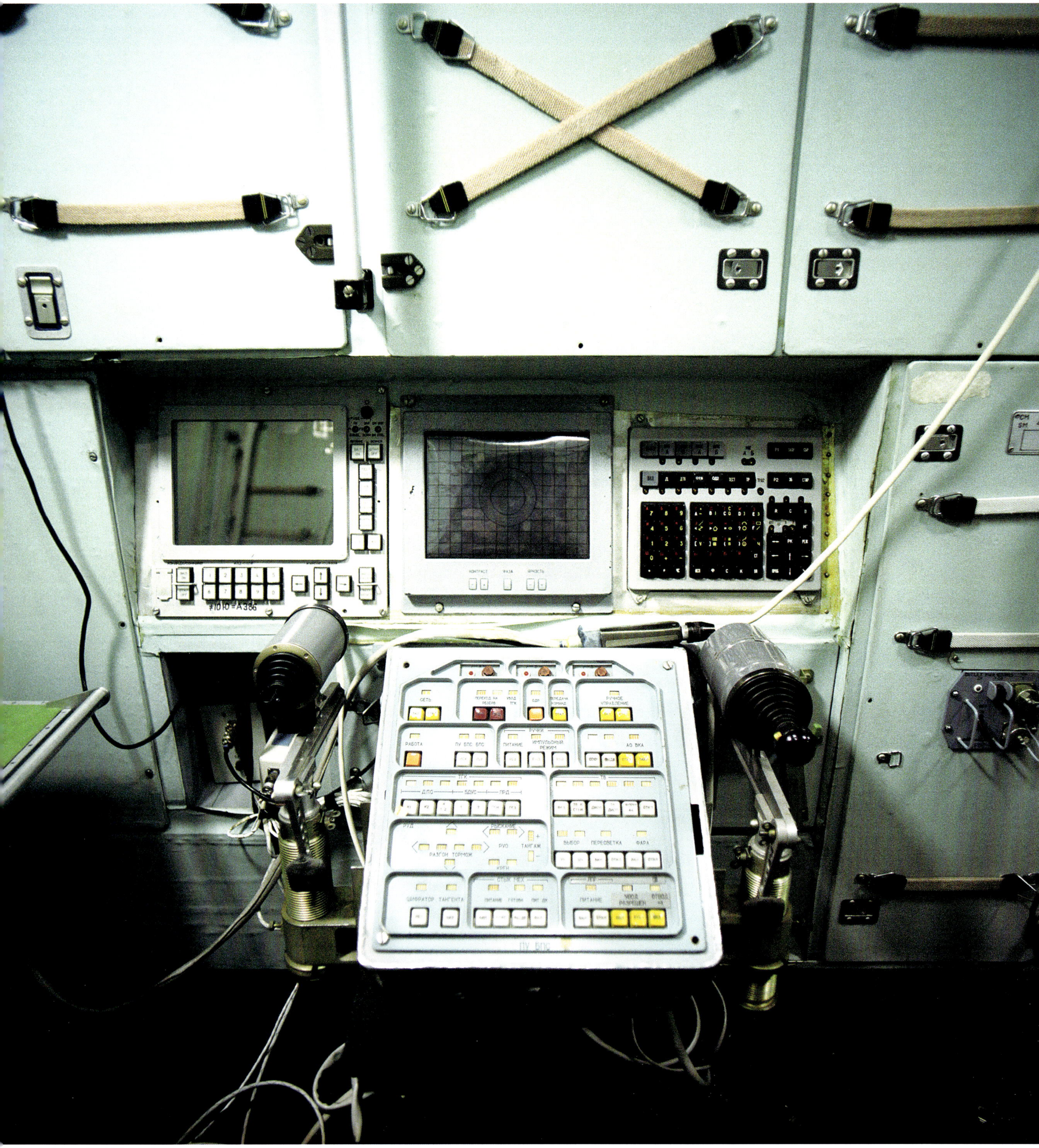

Tele-operator's manual control panel, monitoring and control panel, Mir space station mock-up.

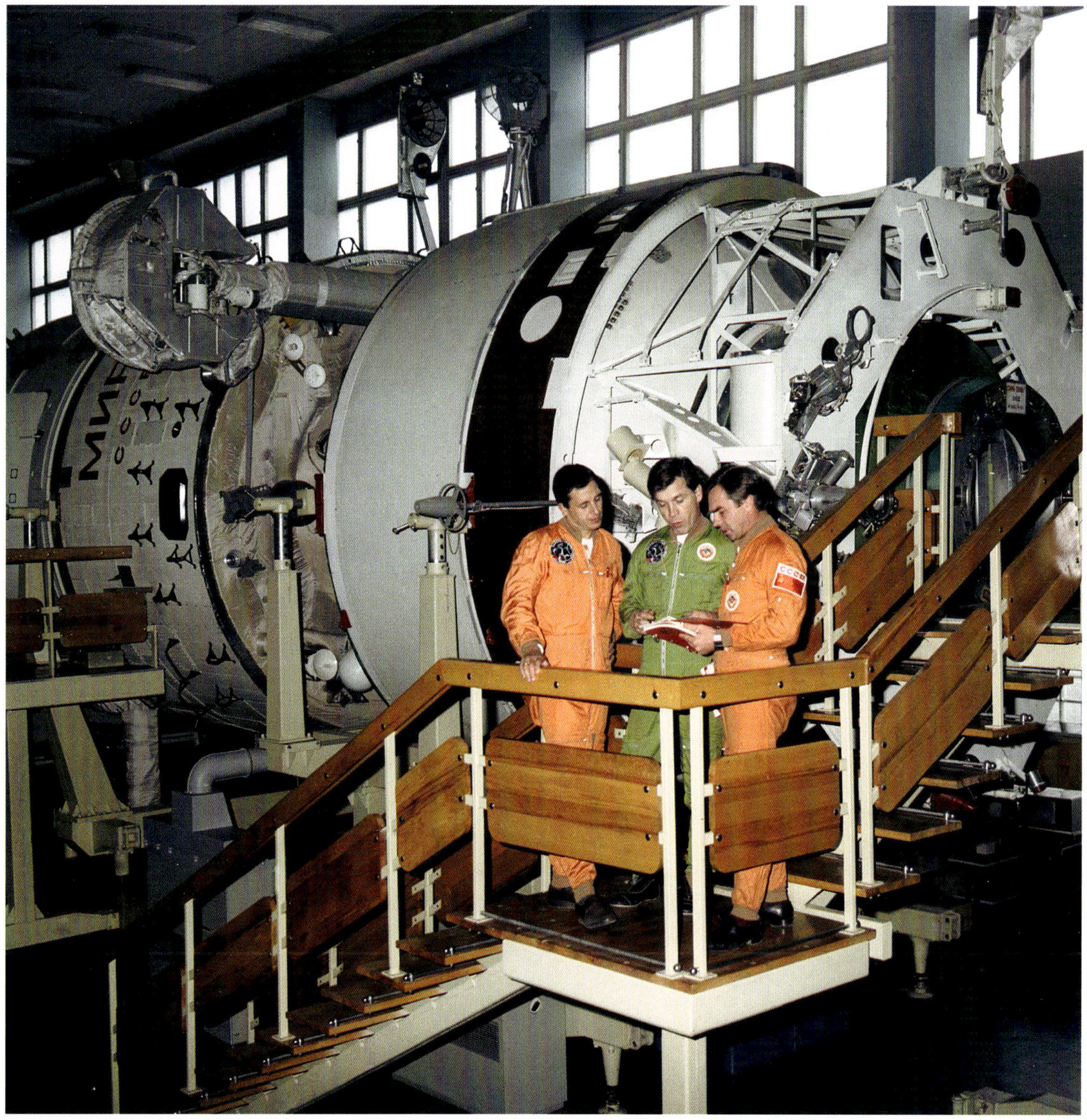

Above: Cosmonauts, A A Volkov, S K Krikalev and Jean-Loup Marie Chrétien, the space flight crew of Soyuz TM–7, launched 26 November 1989.

Right: Mir space station mock-up detail, Star City.

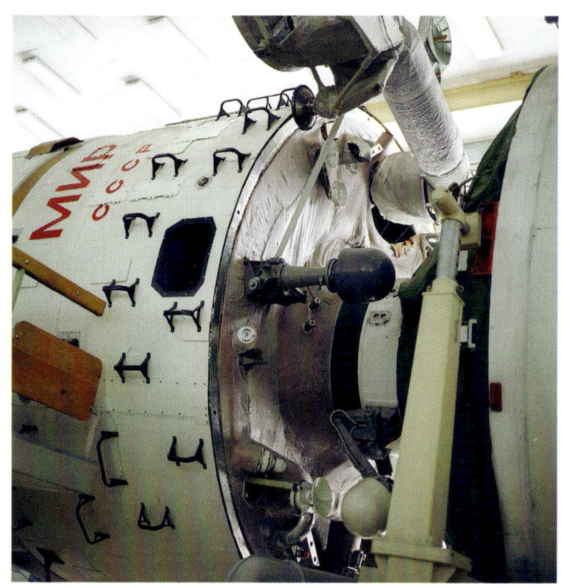

Unused crash dummy.

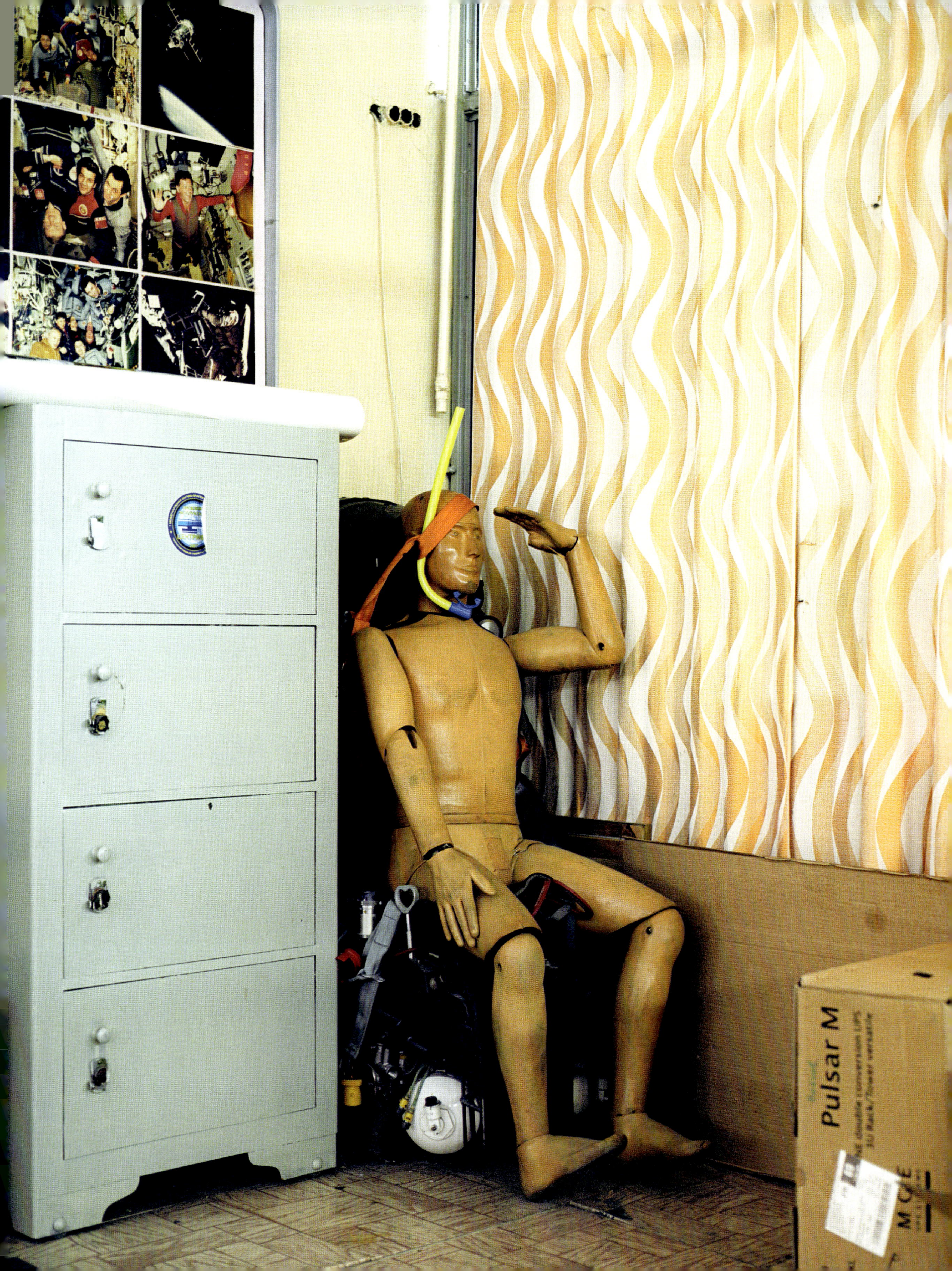
Pulsar M

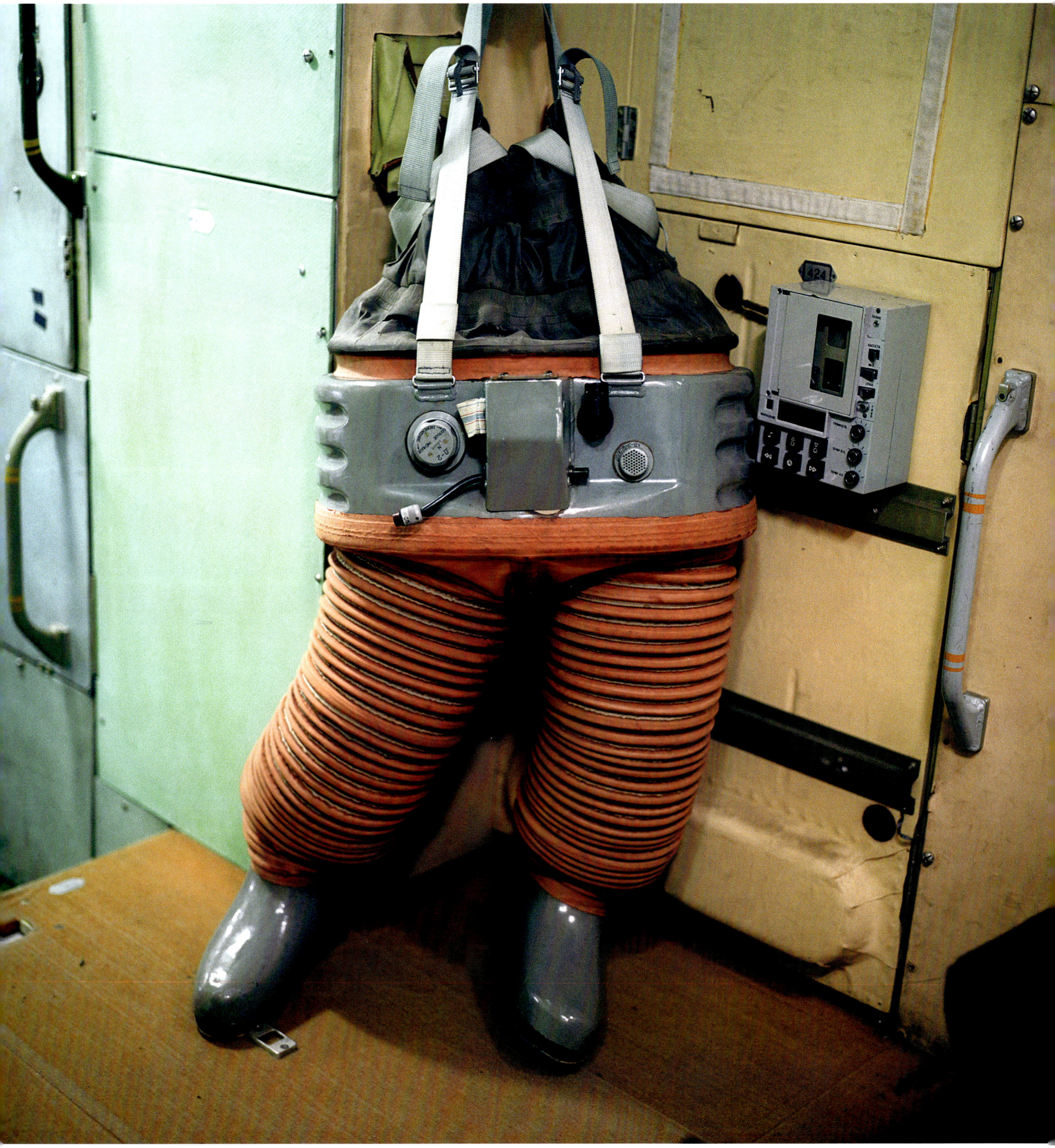

Negative pressure suit (ODNT) used to stabilise blood flow during weightlessness.
It increases the load on the muscles and skeleton, which partially compensates for lack of gravitation.

Bottom left: Cosmonauts, S Y Savitskaya, the second woman in space and the first woman to perform a space walk (25 July 1984), and V A Dzhanibekov and I P Volk, space flight crew of Soyuz T–12, launched 17 July 1984.

Bottom right: 1980 postage stamp commemorating the 20th anniversary of the foundation of the Yuri Gagarin Training Centre, illustrating one of the cosmonauts wearing negative pressure suit (ODNT).

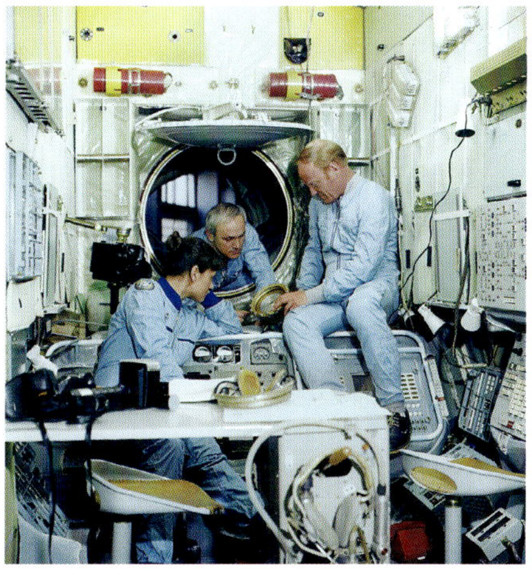

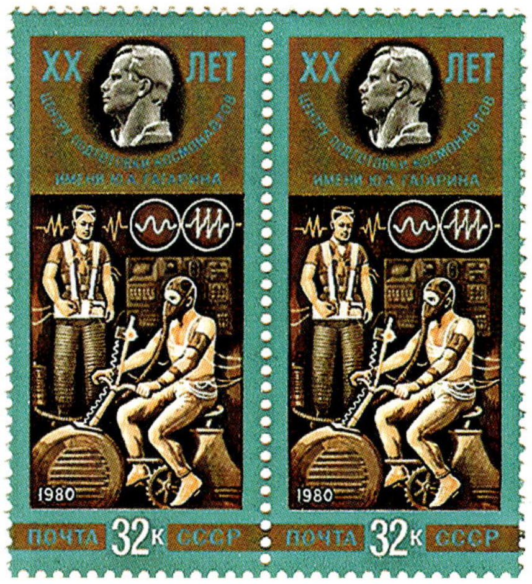

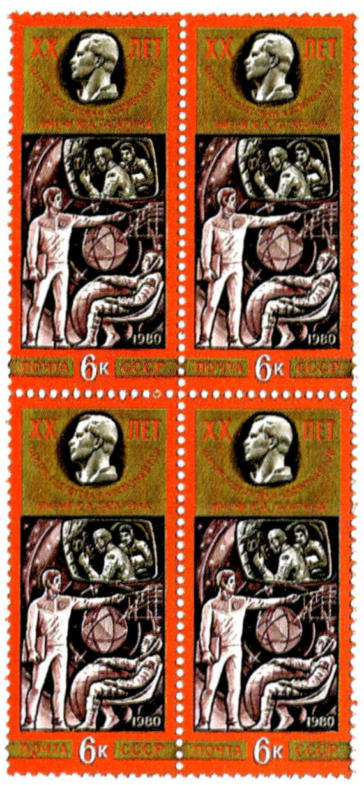

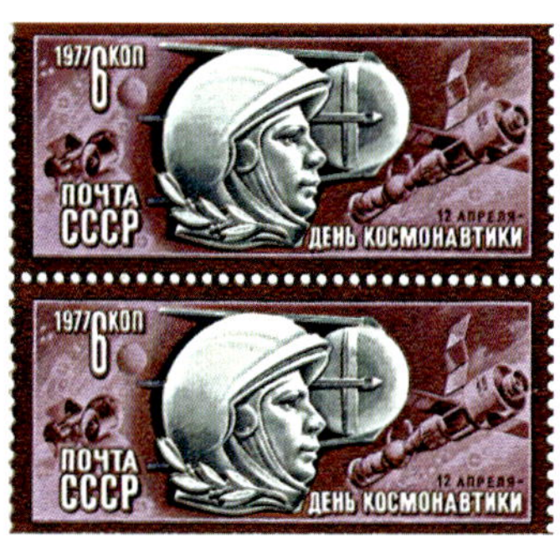

Left: Postage stamp (1980) commemorating the 20th anniversary of the foundation of the Yuri Gagarin Training Centre, illustrating cosmonauts at control panels.

Right: Postage stamp (1977) commemorating Cosmonautics Day, celebrated each year on 12 April in honour of the first space flight made on 12 April 1961 by Y A Gagarin.

Below: Control room of the TDK-7ST training facility, integrated simulator of spacecraft Soyuz-TM.

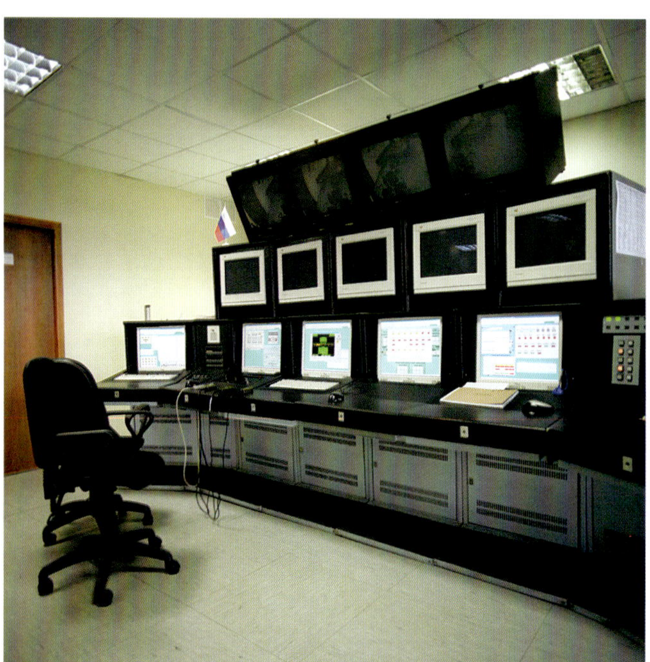

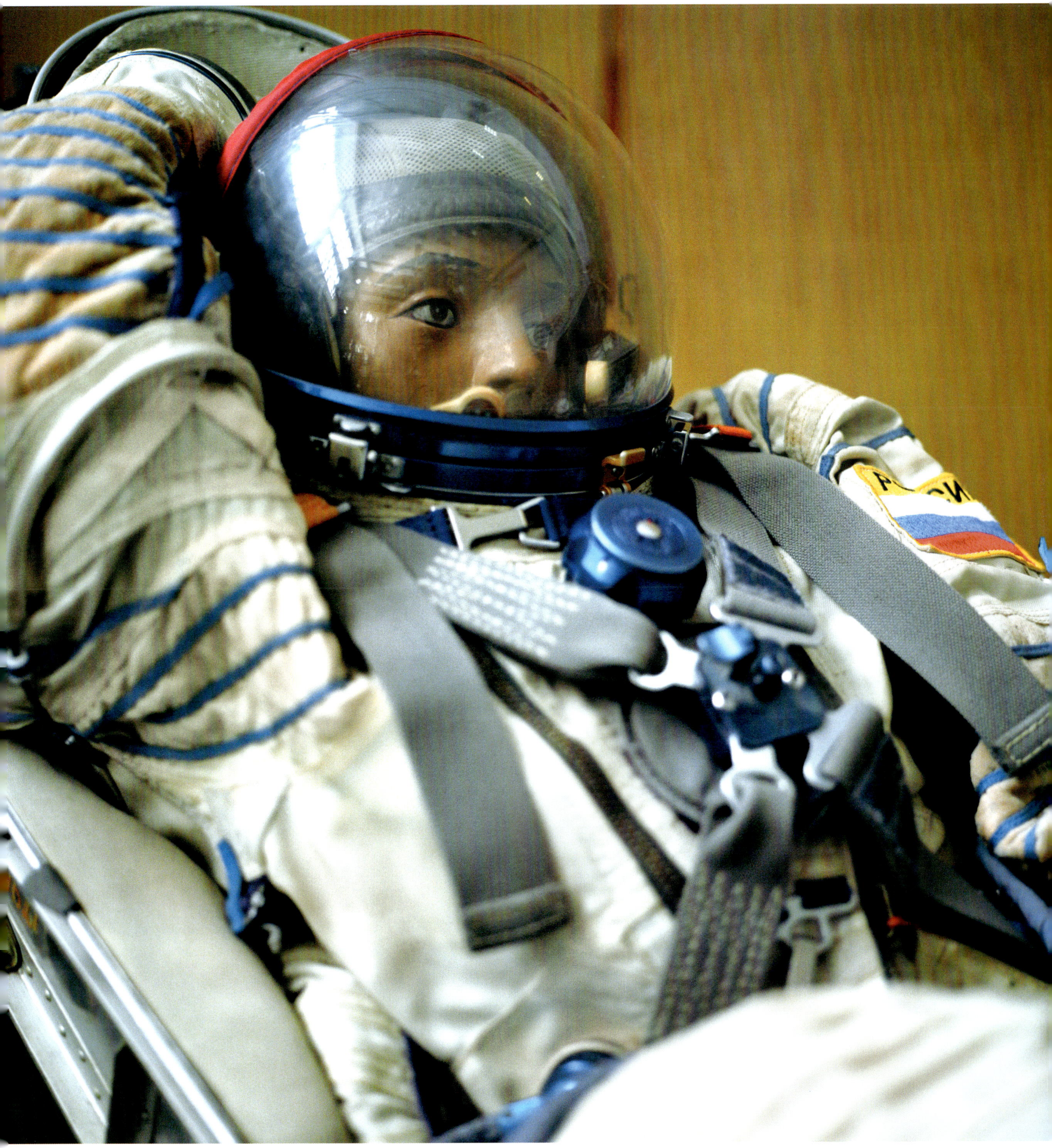

Dummy wearing outdated space suit.

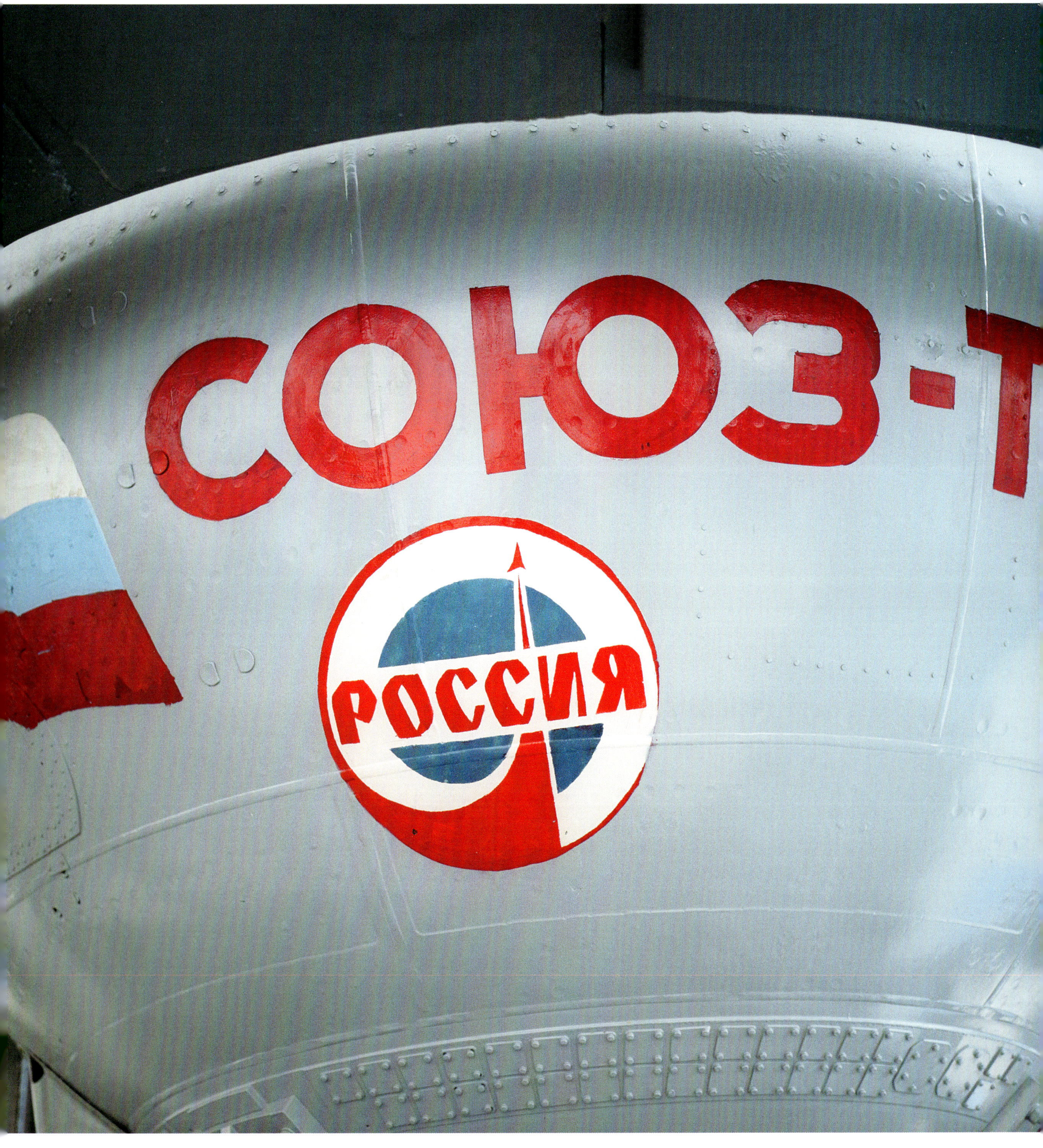

Soyuz–TM simulator.

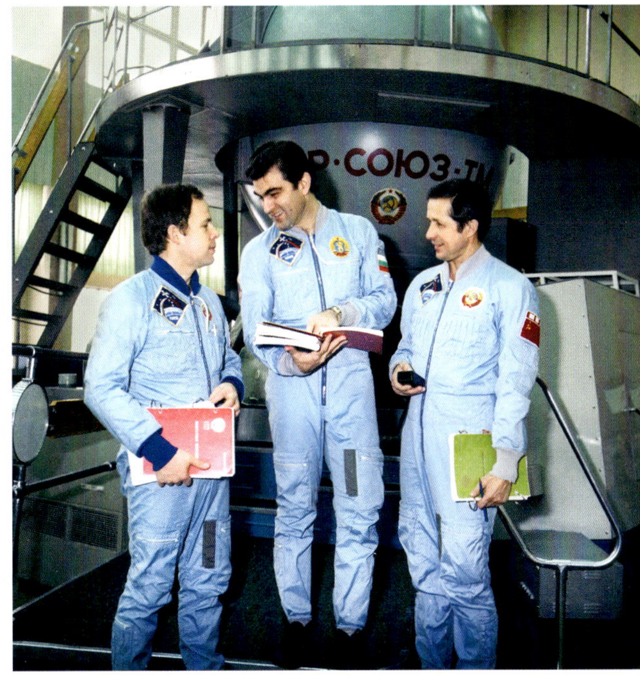

Right: Cosmonauts, A V Solovyov, V P Savinykh and A P Aleksandrov, space flight crew of Soyuz TM–5, launched 7 June 1988, in front of the Soyuz–TM simulator, which is still used to train cosmonauts.

Below: Control panel, inside the Soyuz–TM simulator.

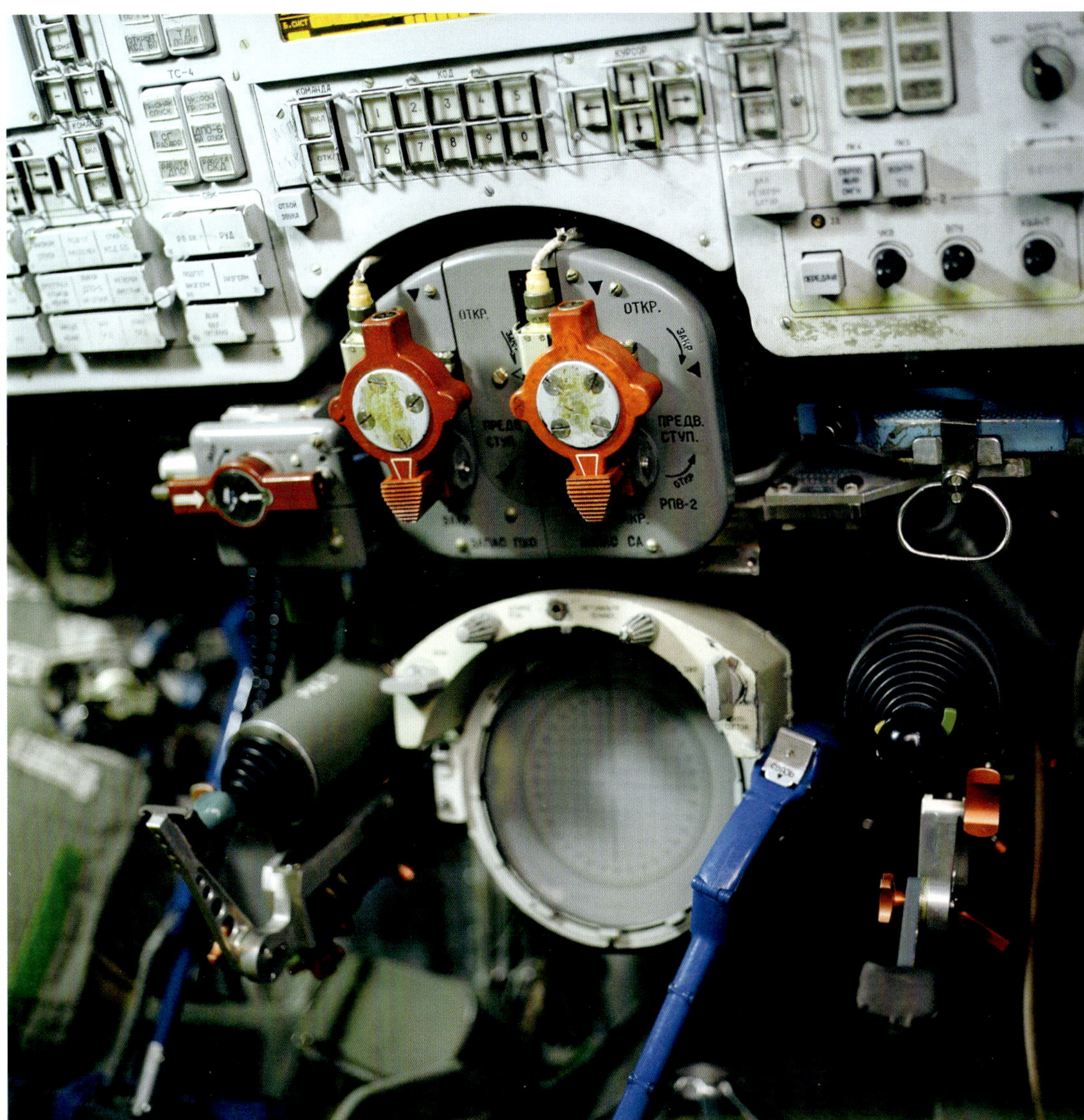

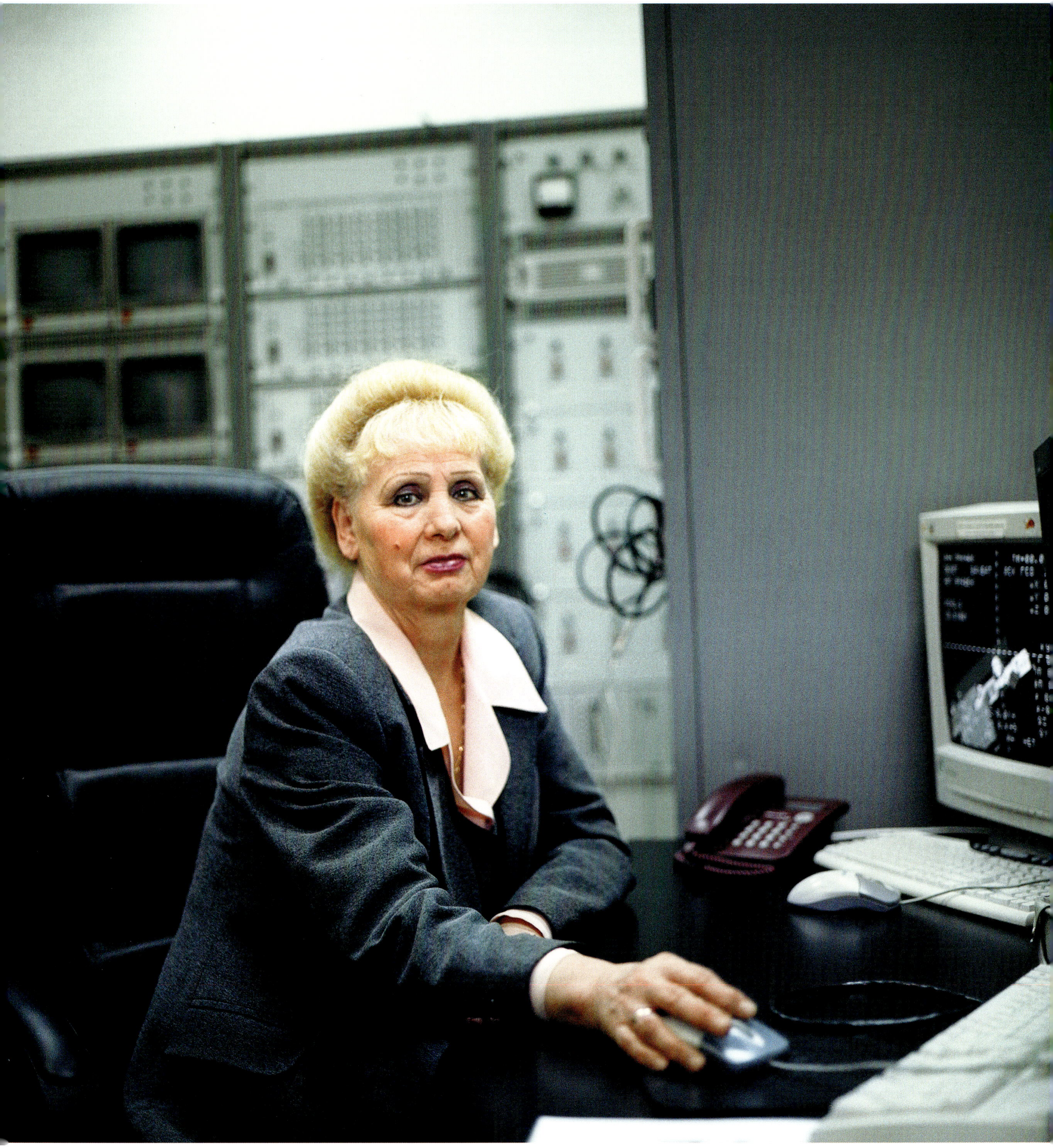

Engineer, A Osipova at the control panel of the TDK–7ST simulator intended for crew preparation for flight on Soyuz–TMA spacecraft.

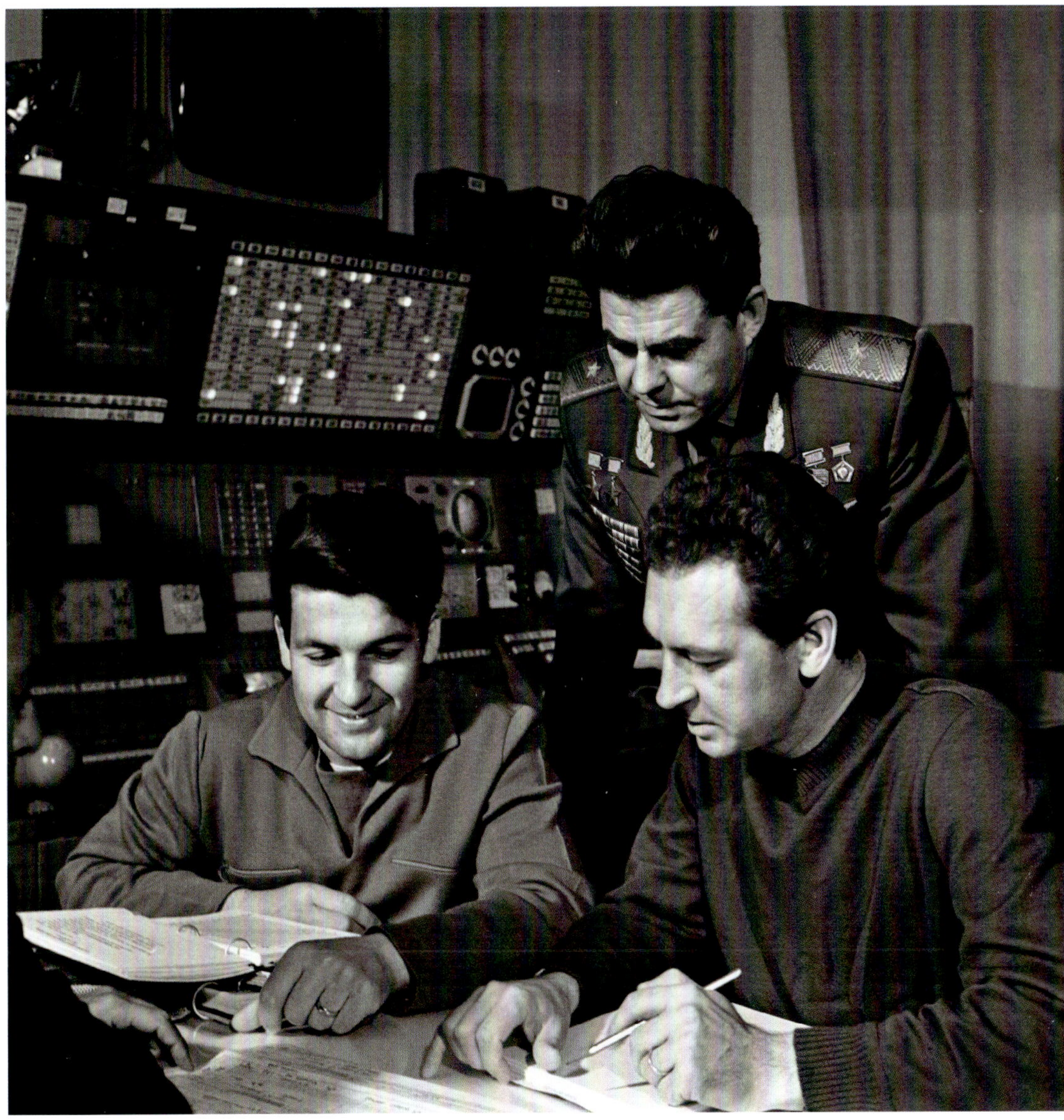

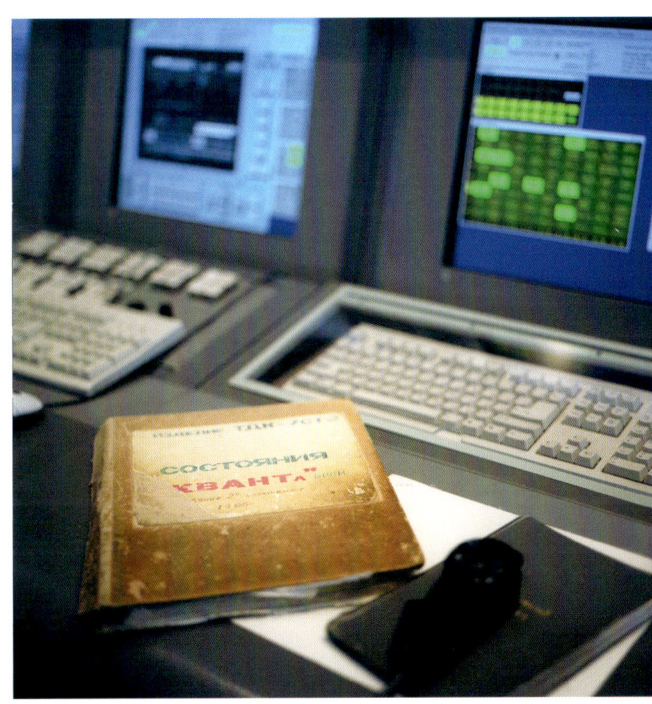

Above: Cosmonauts, P I Klimuk and V V Lebedev, Soyuz–13 space flight crew, launched 18 December 1973.

Right: Control panel of TDK–7ST simulator.

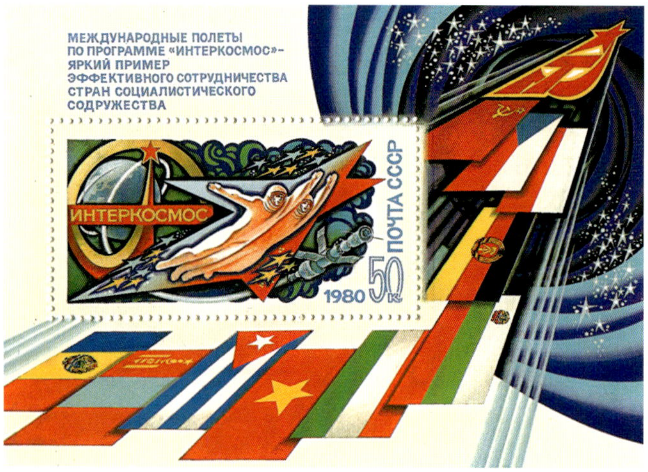

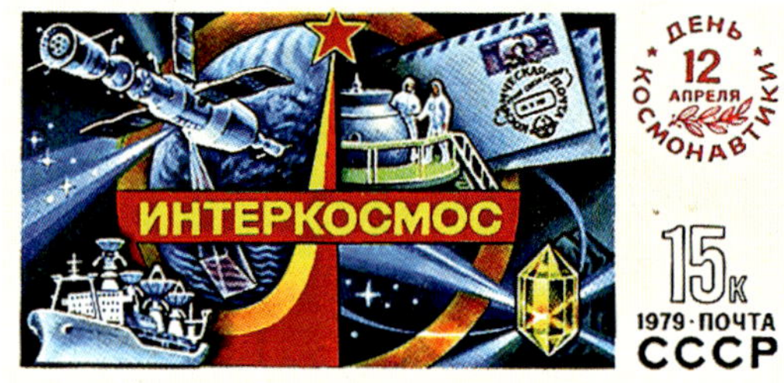

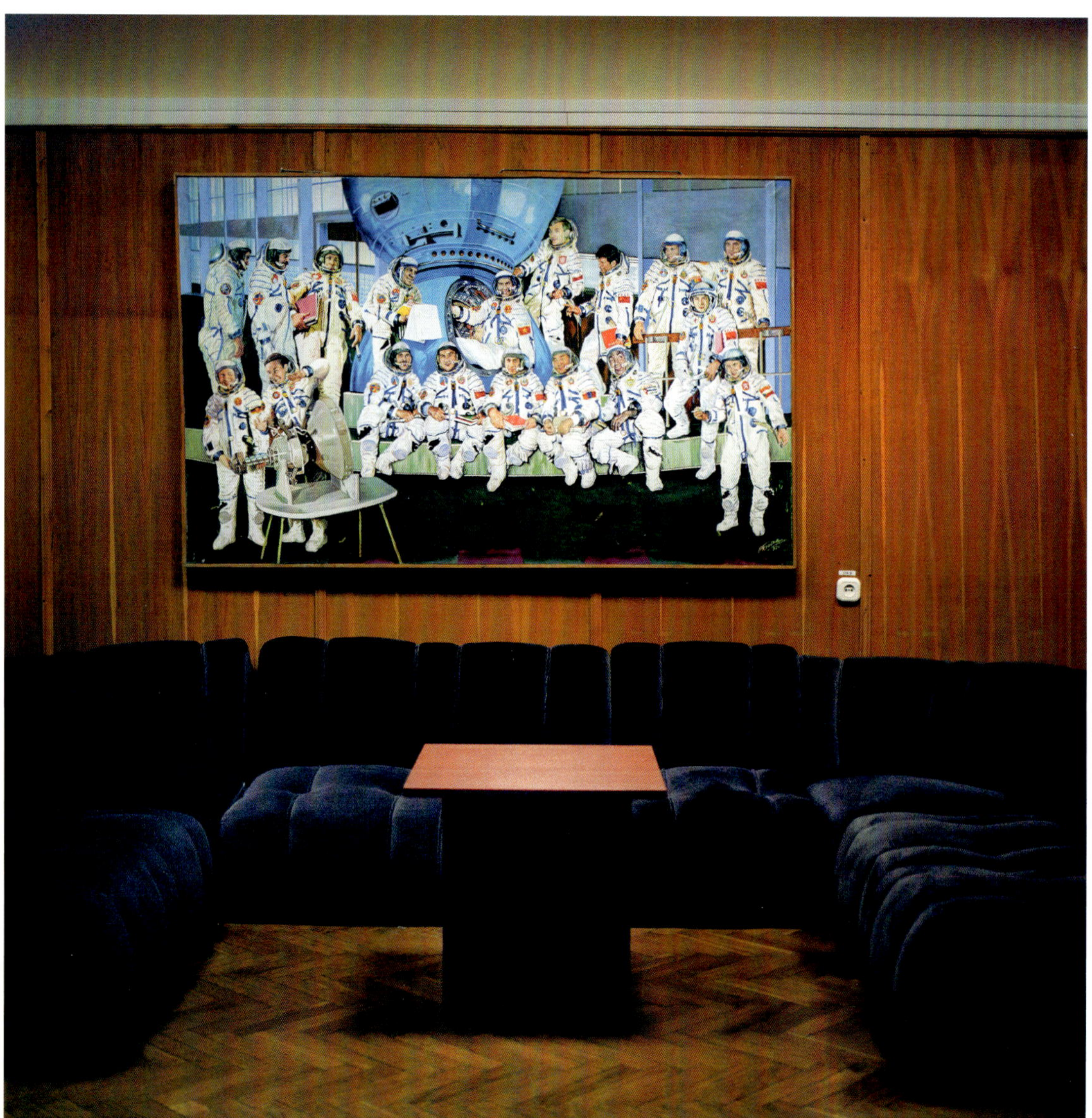

Above: Postage stamps (1980) commemorating Interkosmos space exploration programme run by the Soviet Union to allow members from military forces of allied Warsaw Pact countries and also some sympathetic countries, such as France and India, to fly into space.

Left: Painting featuring members of Interkosmos space exploration programme, by R Bobovich, 1982. Reception area, 2nd Department, Star City.

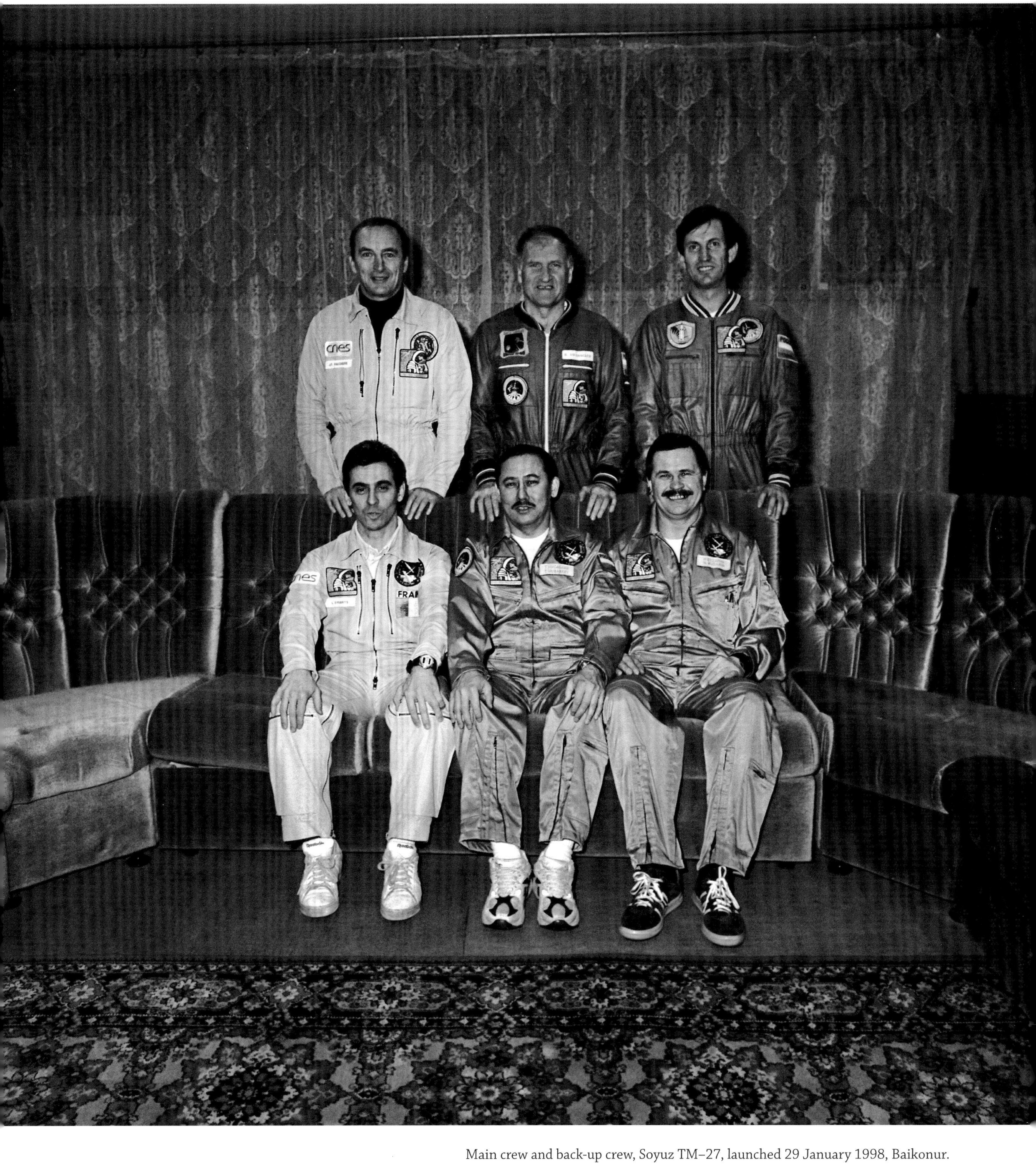

Main crew and back-up crew, Soyuz TM–27, launched 29 January 1998, Baikonur.

An envelope commemorating the allied Russian / Polish Soyuz–30 space flight (1978), signed by cosmonauts P R Popovich, A A Leonov, the first man to conduct a space walk, V F Bykovsky and V V Gorbatko.

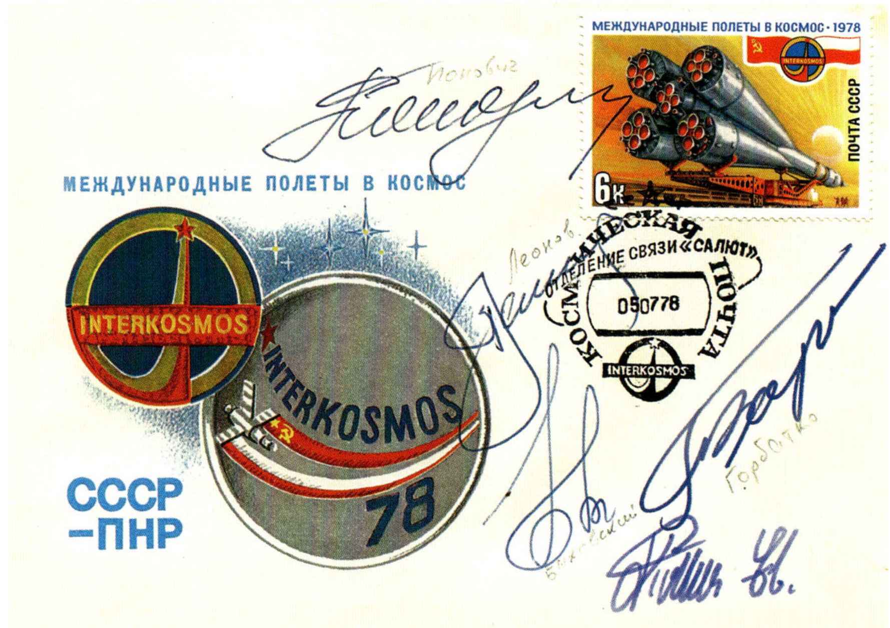

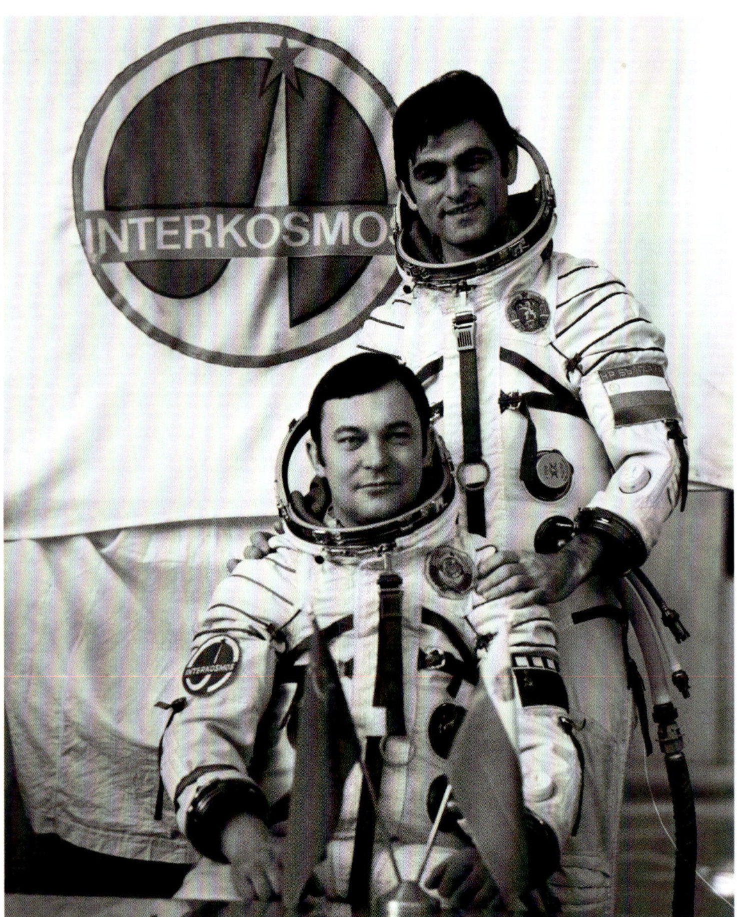

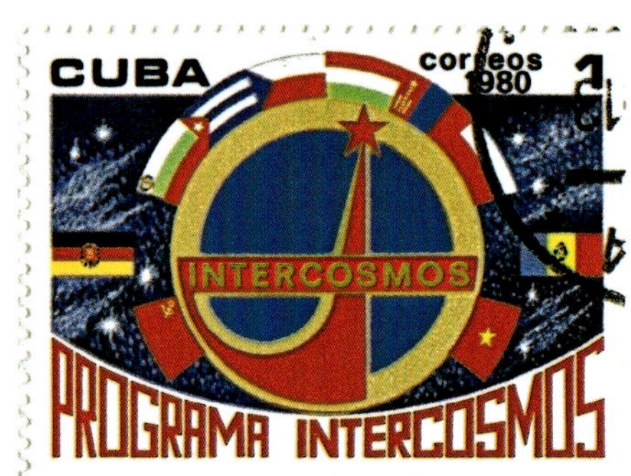

Above: Cuban postage stamp commemorating the Interkosmos space exploration programme.

Left: Cosmonauts, Y V Romanenko and Arnaldo Tamayo Mendez, Soyuz–38 space flight crew, launched 18 September 1980.

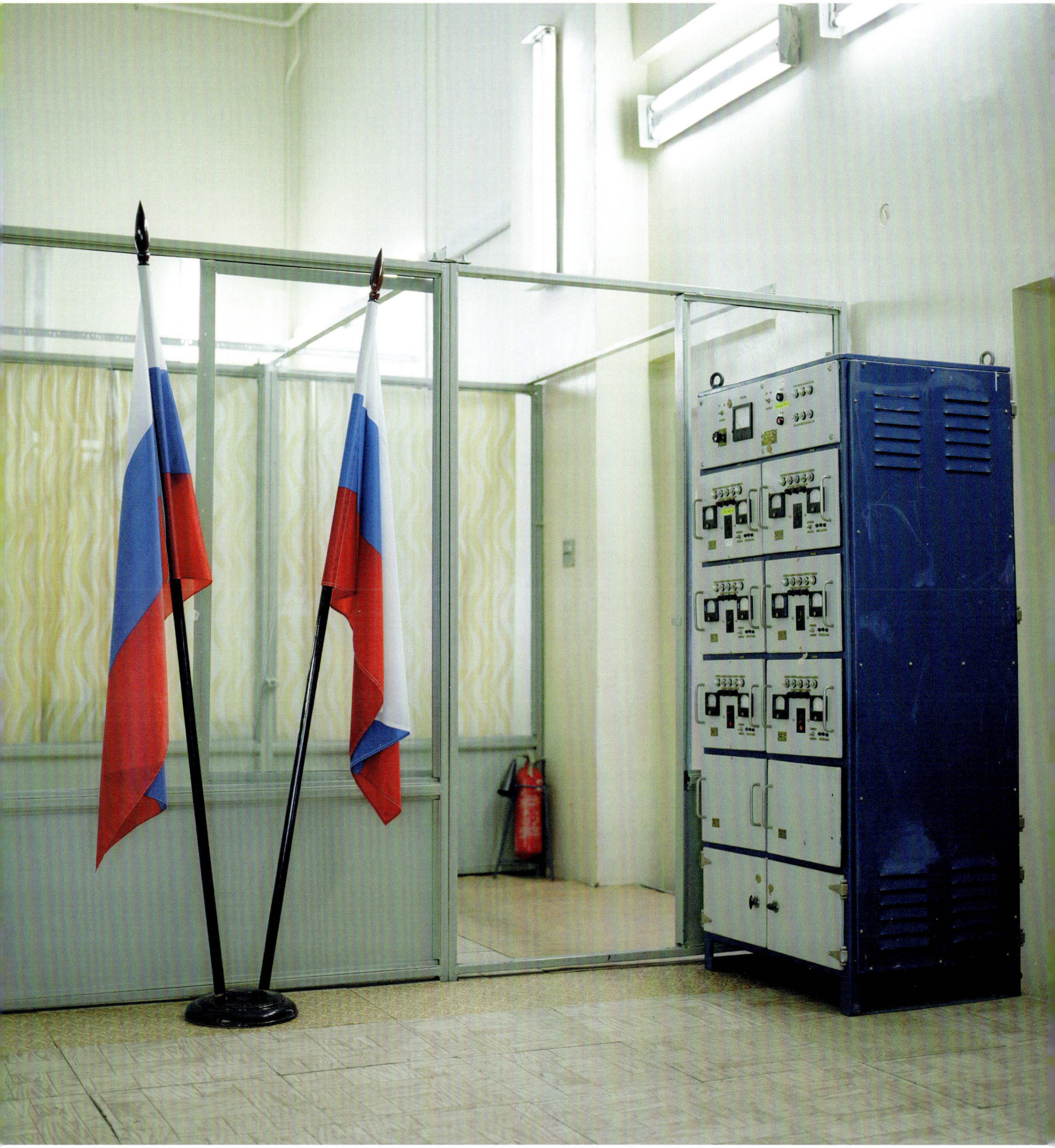

Redundant fuse panel.

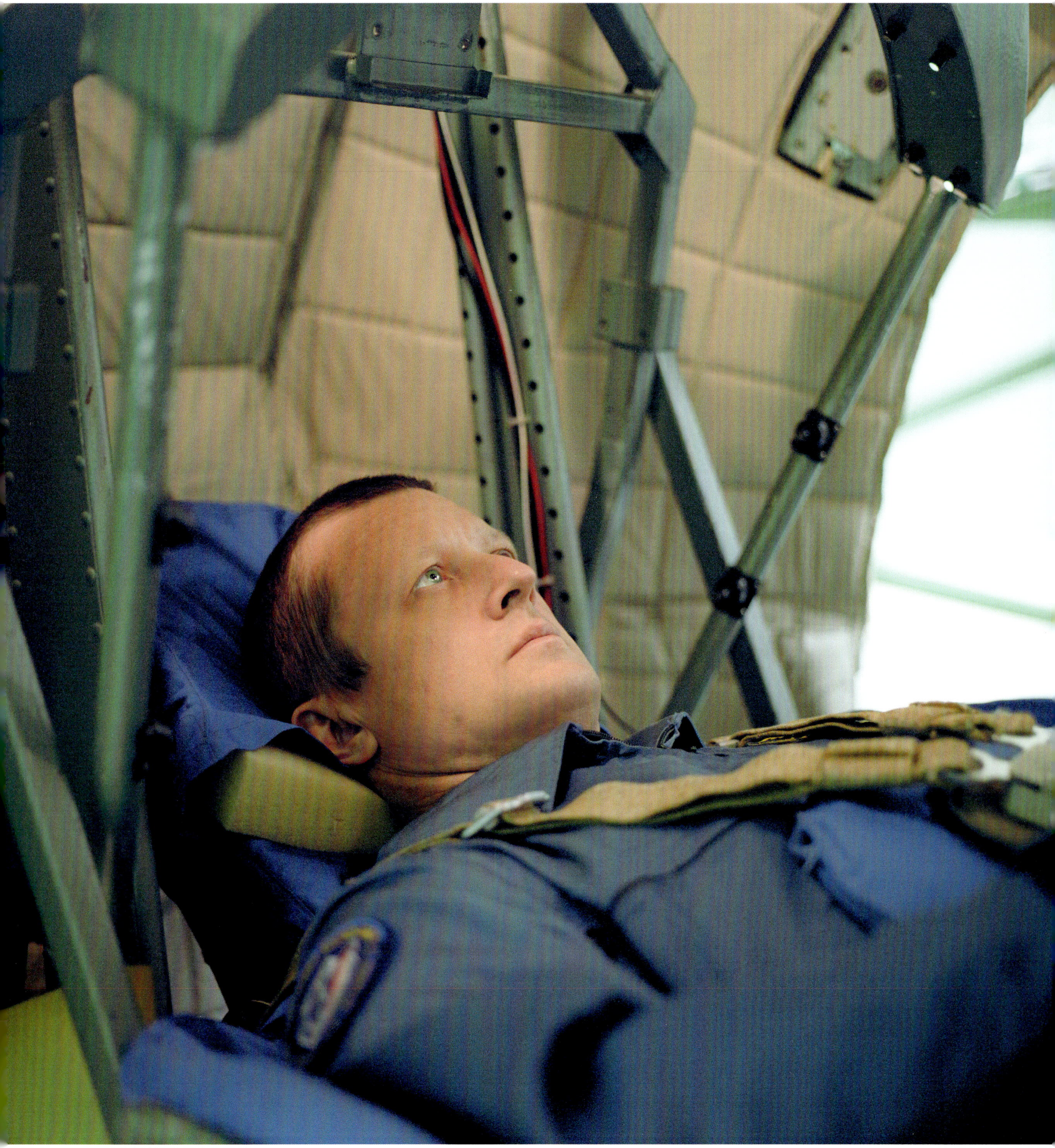

Cosmonaut, D Y Kondratyev at a training session, TsF–7 centrifuge.

Far right: 1981 postage stamp commemorating the Interkosmos Soyuz–40 space flight accomplished by Soviet cosmonaut, L I Popov and Romanian astronaut, Dumitru Prunariu.

Right: TsF–7 centrifuge in motion, designed to check the g-load tolerance.

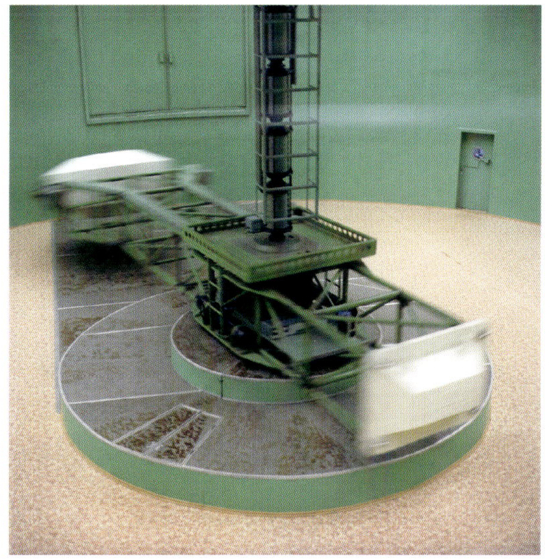

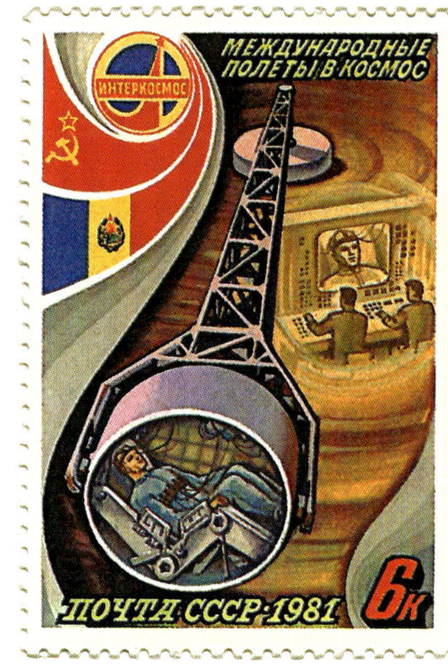

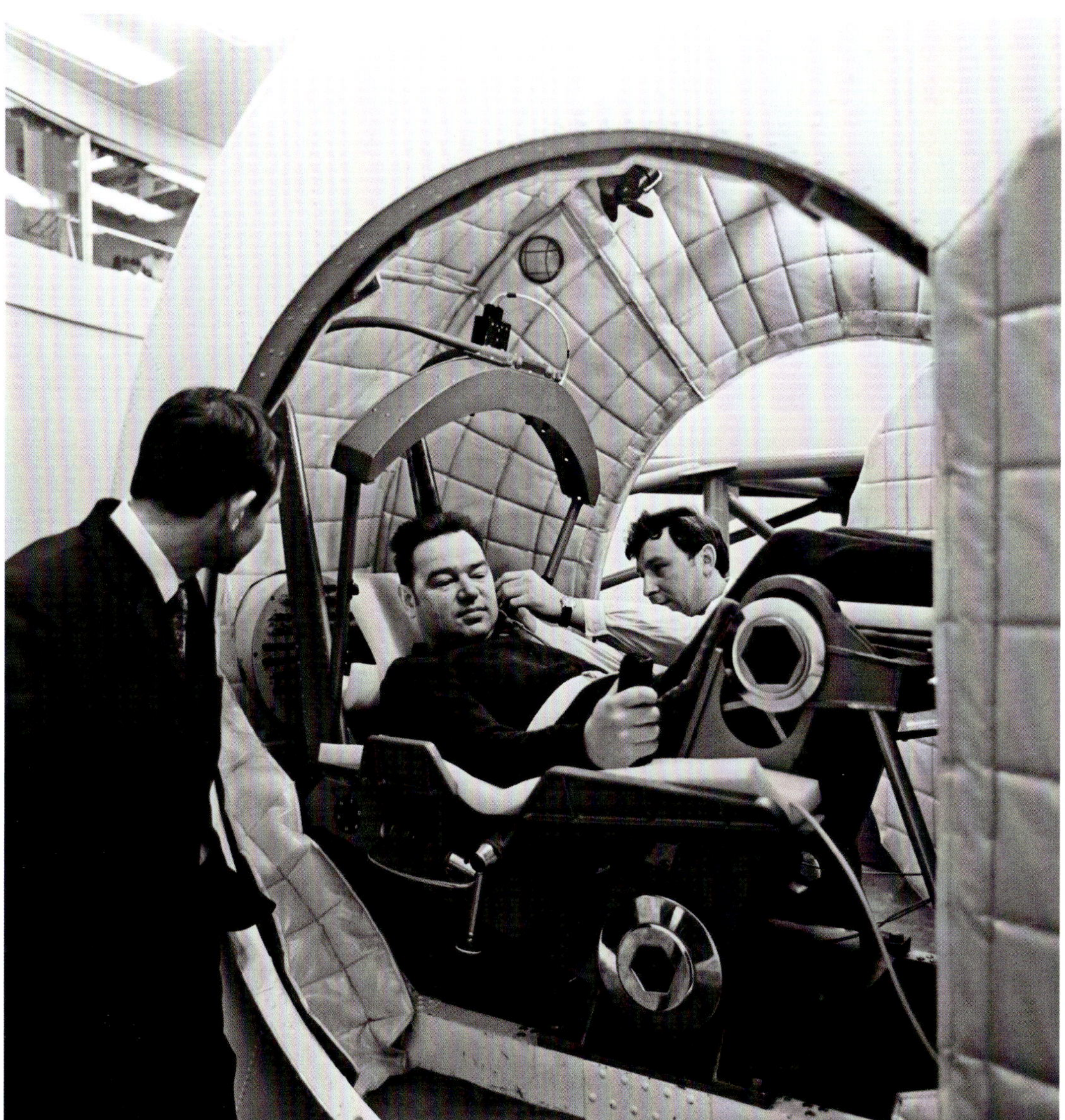

Cosmonaut, G M Grechko getting ready for his test on the TsF–7 centrifuge.

Techical features of the TsF–7 dynamic trainer centrifuge:

- rotation radius 7m
- max load 20g
- max load gaining rate 7g/sec
- motor capacity
  nominal 820kW
  starting 1280kW

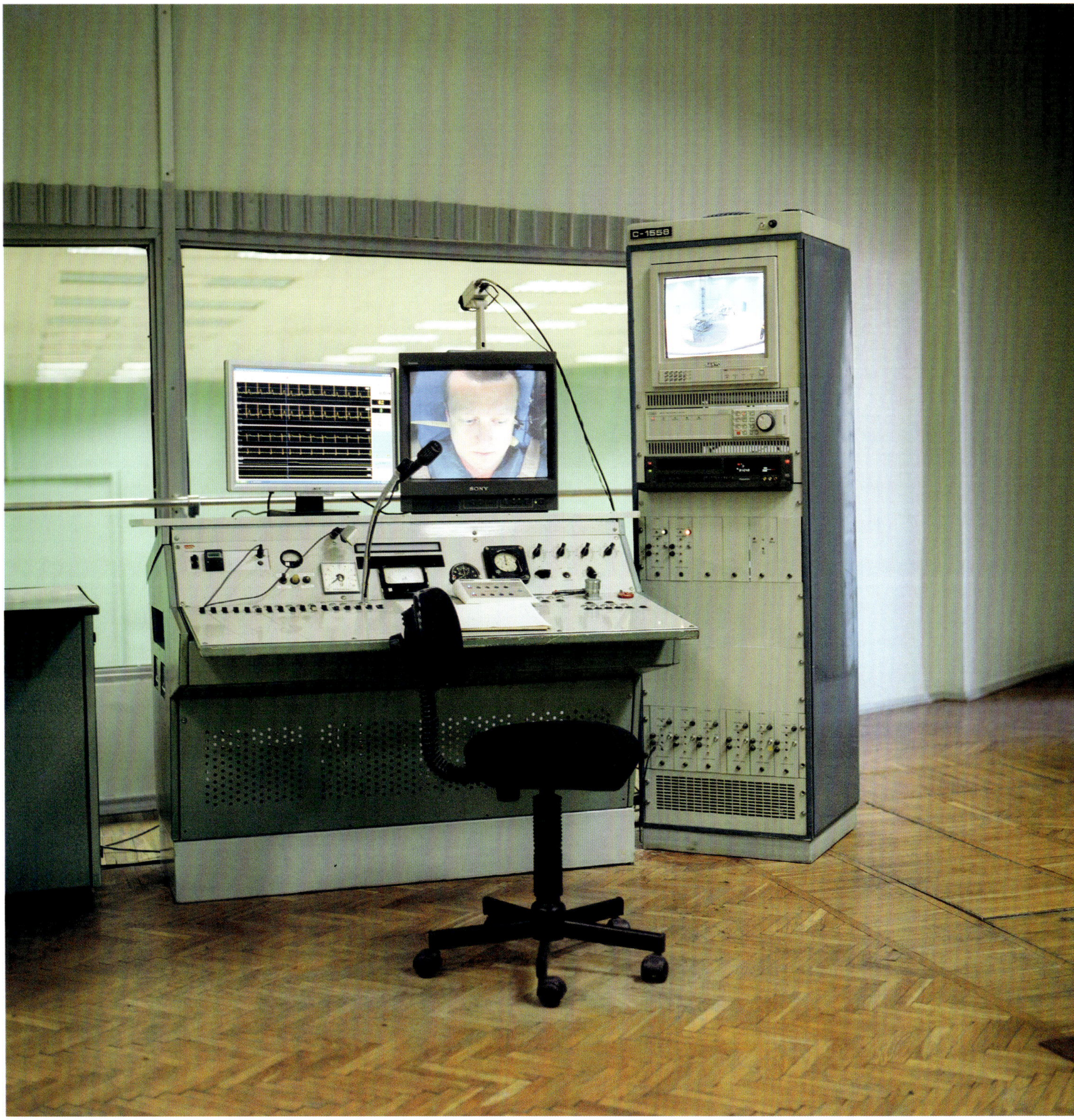

Medical control panel, dynamic trainer TsF–7.

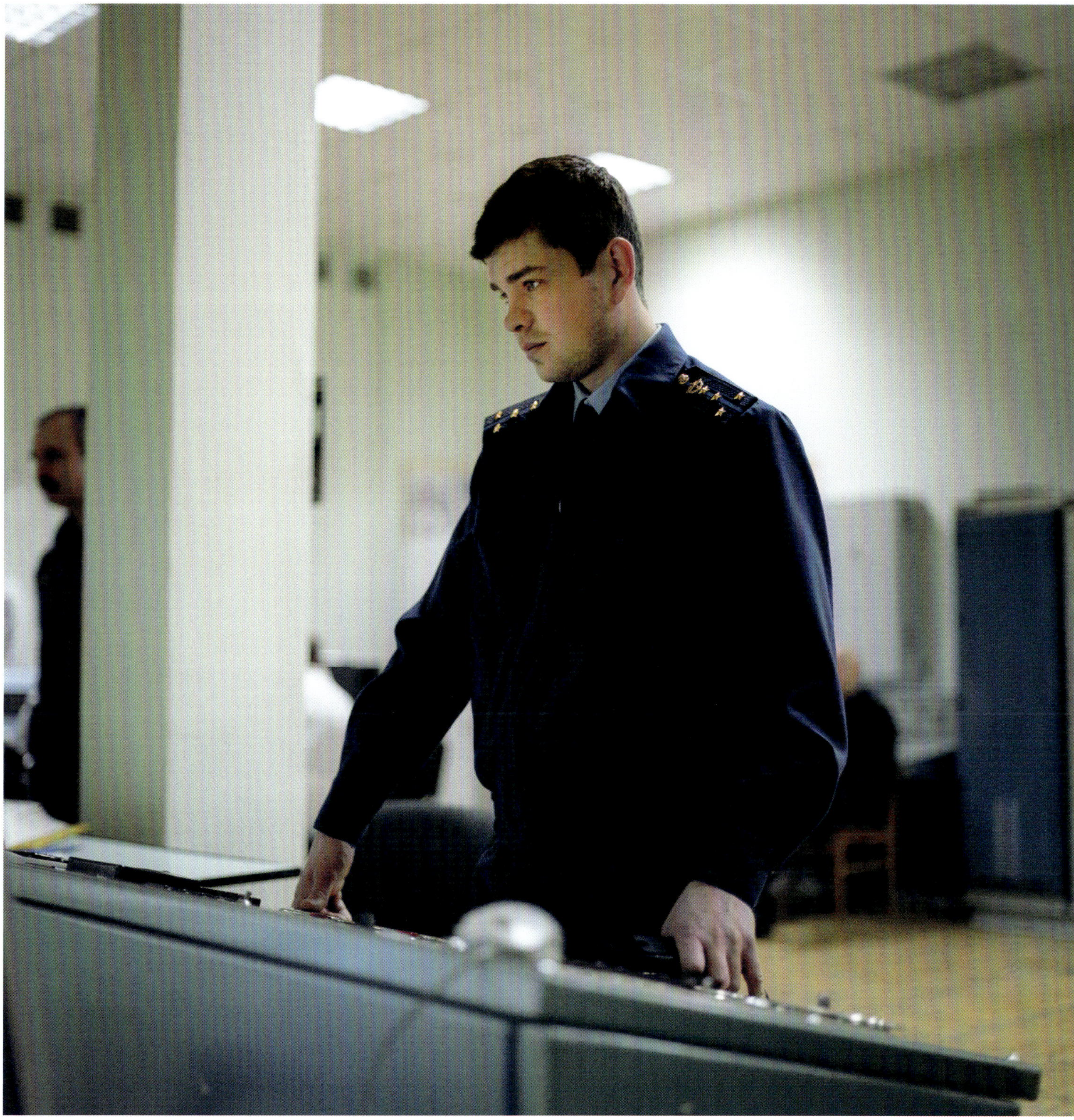

Engineer, S Sergeev at the TsF-7 centrifuge simulator control panel.

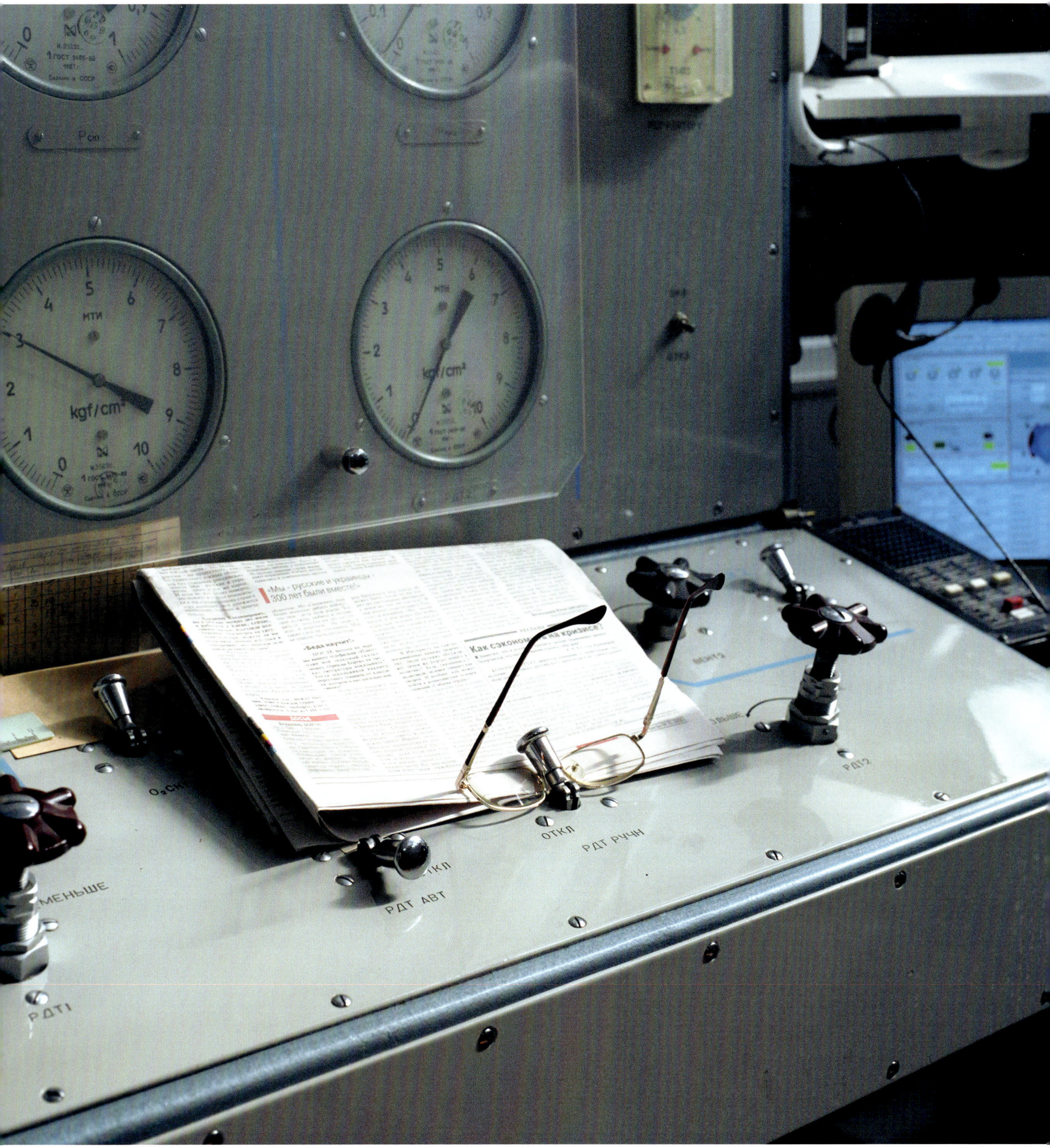

Control panel.

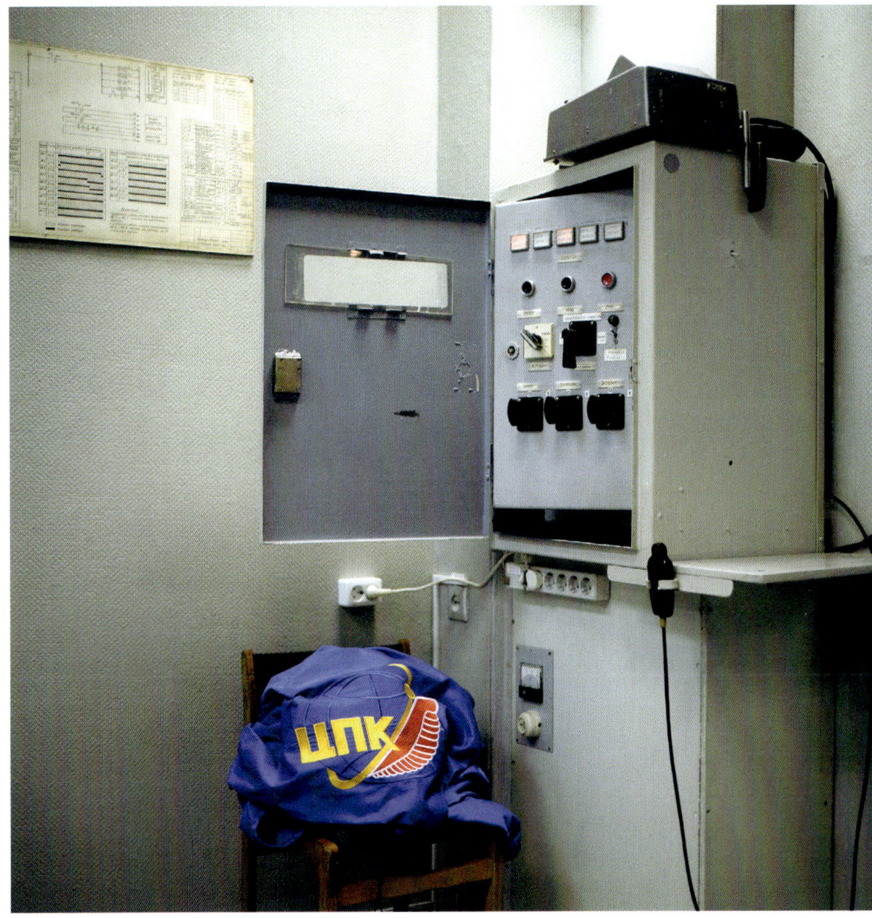

TsF–18 centrifuge simulator's step platform remote control panel.

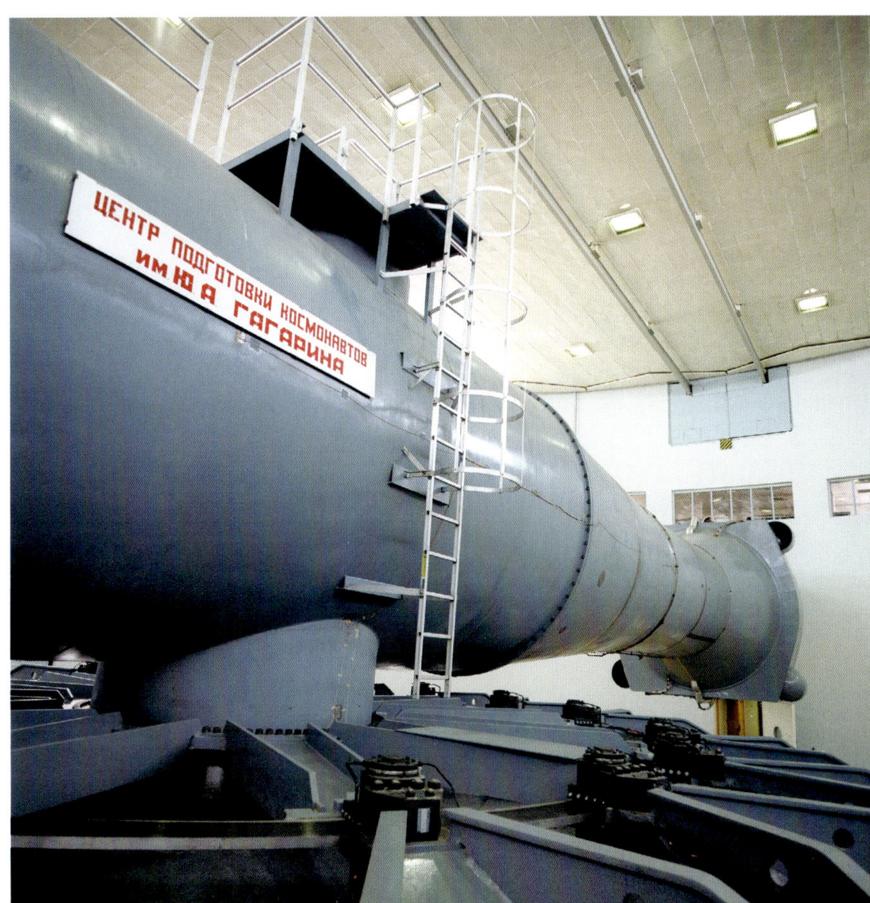

TsF–18 centrifuge simulator used to check the g-load tolerance and to simulate negative elements of space flight, such as longitudinal g-load, physiological microgravity, low cabin pressure, and variable temperature, humidity, and gas composition of the cabin's air.

Techical features of the TsF–18 centrifuge simulator:

- weight 300t
- shoulder radius 18m
- max load 20g
- max load gaining rate 5g/sec
- starting capacity of the primary electrical drive 28mw
- pressure range 800 to 40mmhg/min
- temperature range 5°c to 50°c
- humidity range 30 to 70%
- gas composition range nitrogen 56 to 83% oxygen 17 to 40% carbon dioxide 0 to 4%

Cosmonaut, A N Shkaplerov, with his hydro-suit.

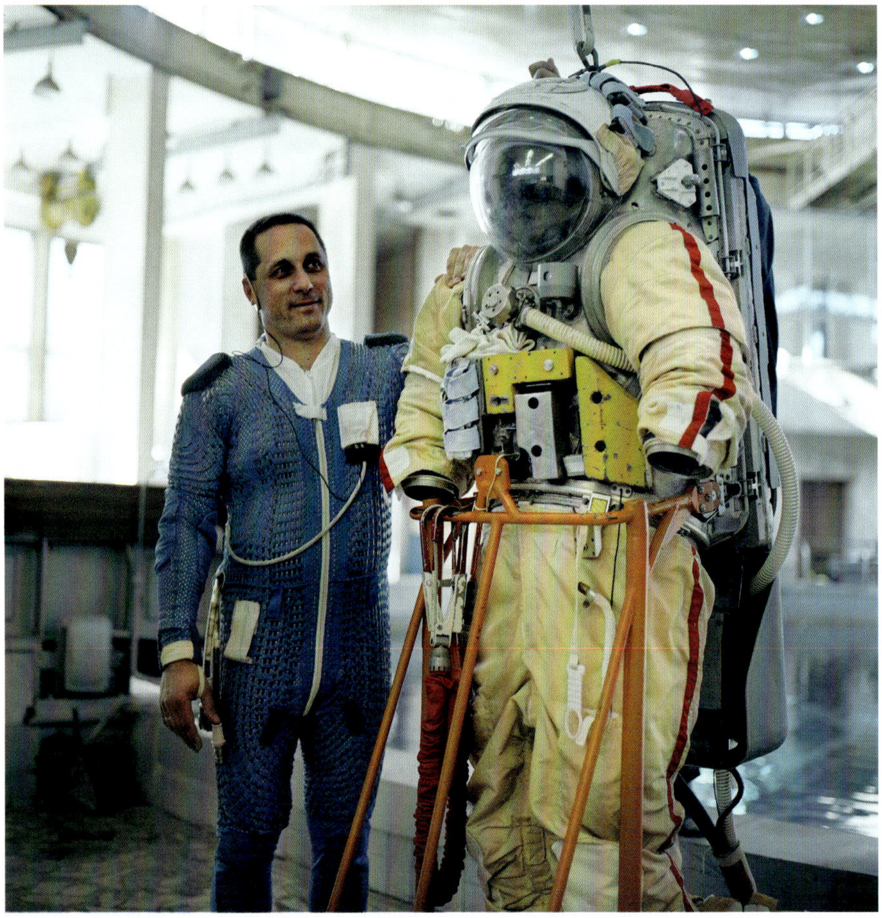

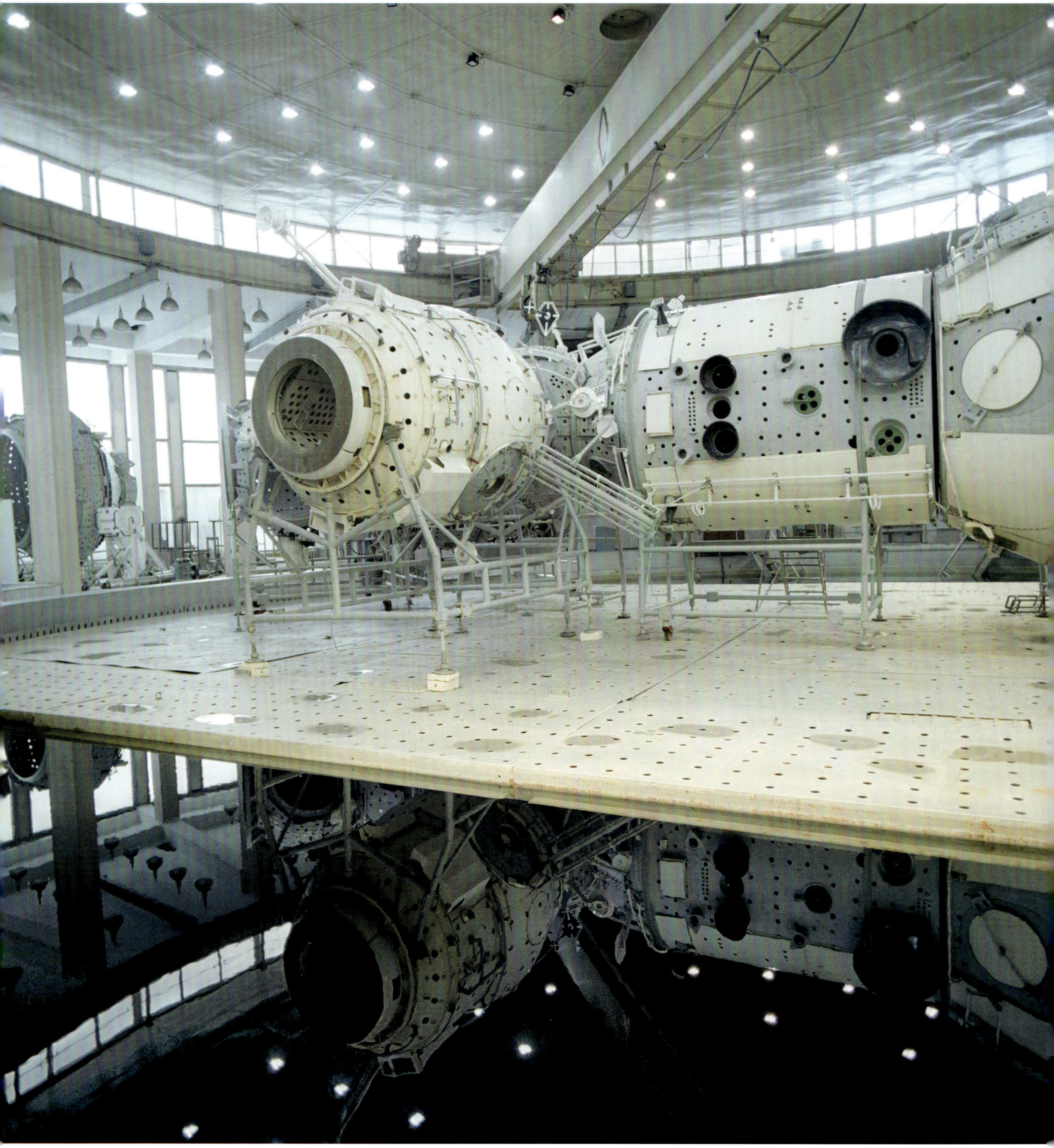

ISS mock-up and pool used for simulating microgravity and for practising extra-vehicular activity in space.

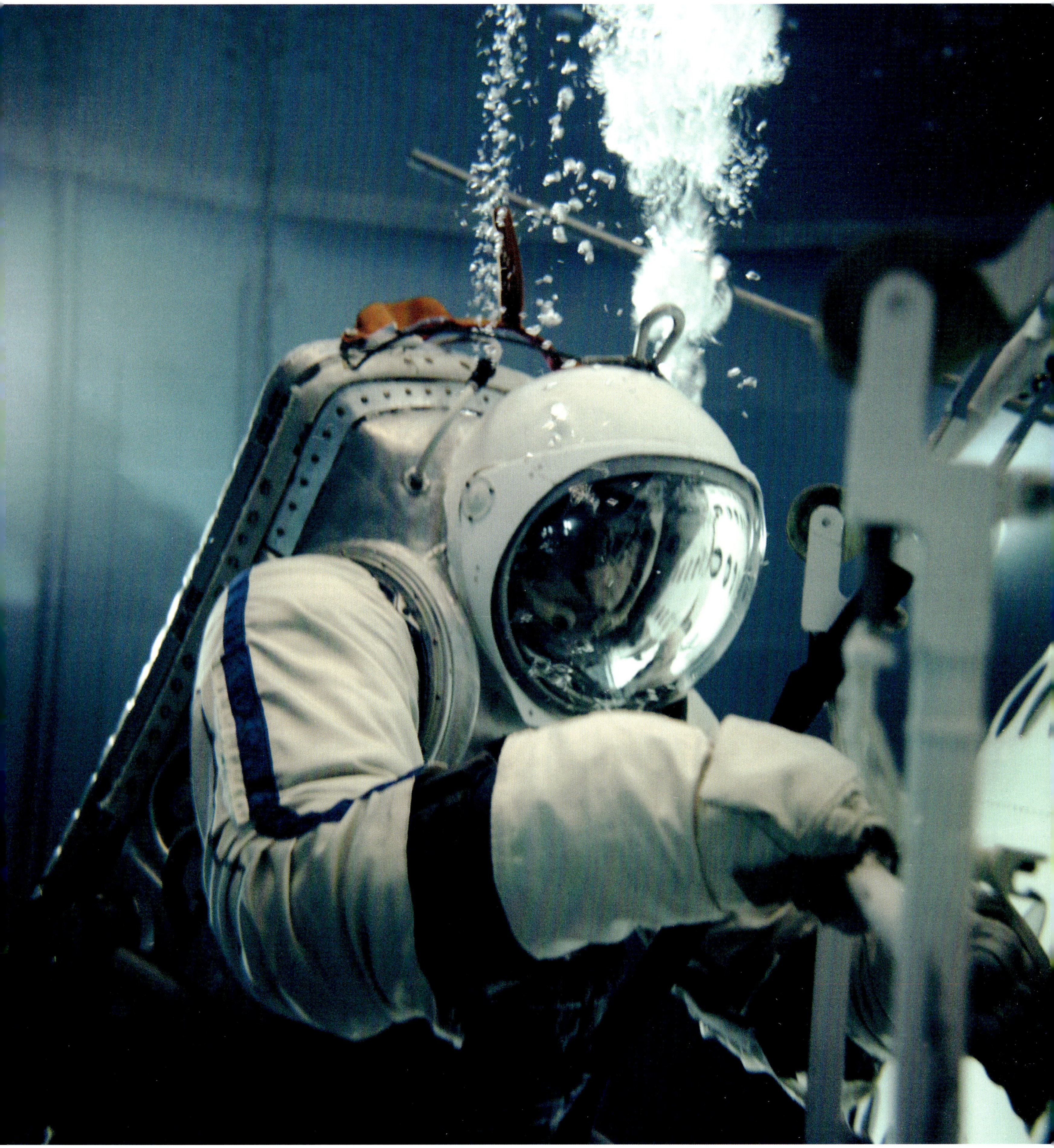

Training session, 1984.

Right: 1980 postage stamp commemorating the 20th anniversary of the foundation of the Yuri Gagarin Training Centre, illustrating cosmonauts training at the Hydro laboratory.

Below: Cosmonauts, A S Viktorenko and A A Serebrov getting ready for their training session at the Hydro laboratory.

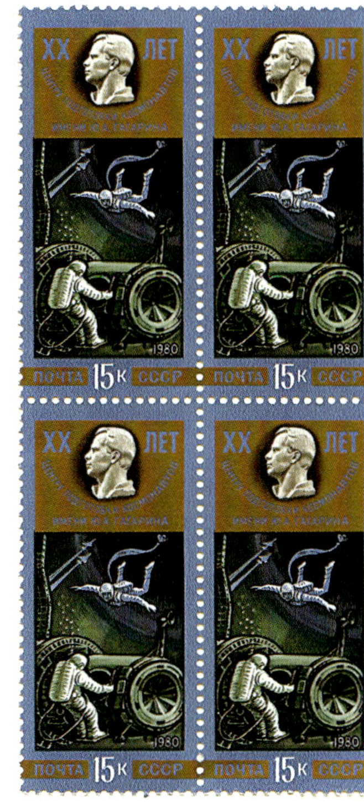

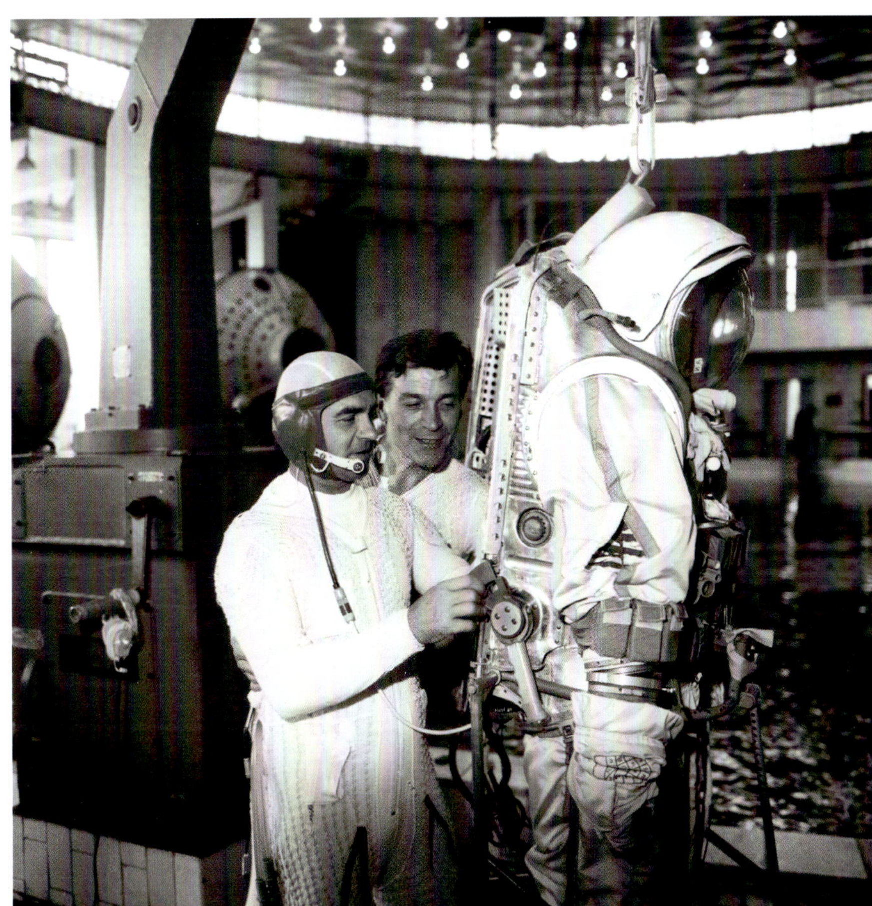

Medic, A Krasnov.

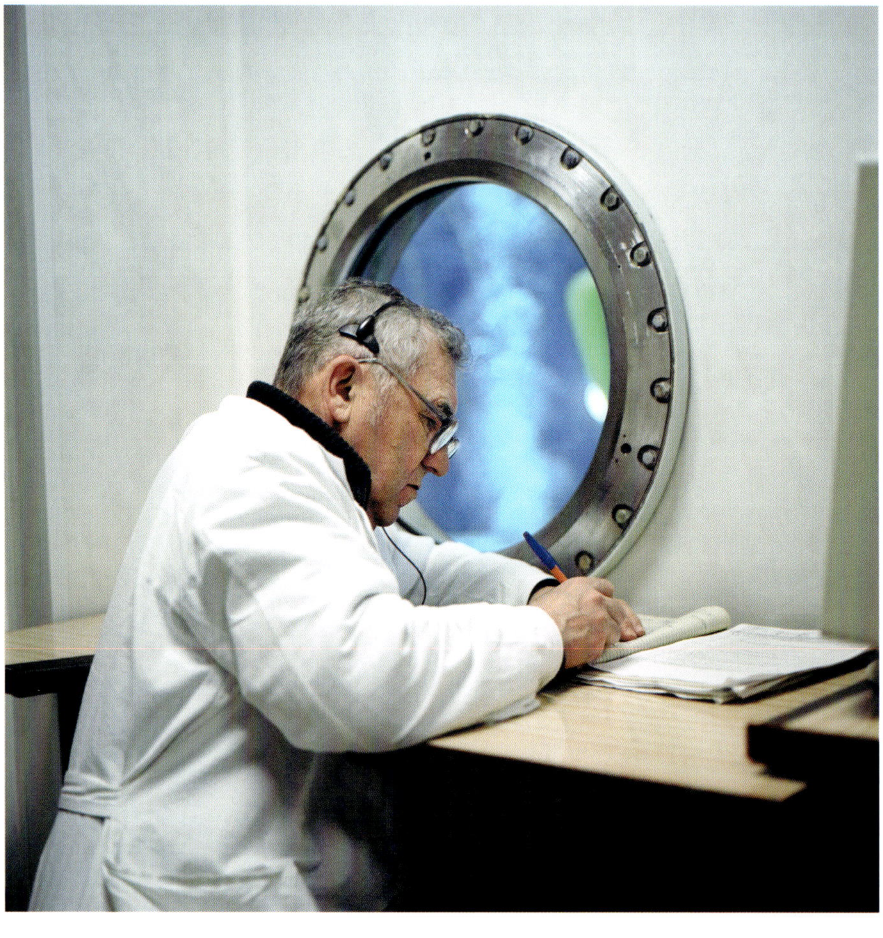

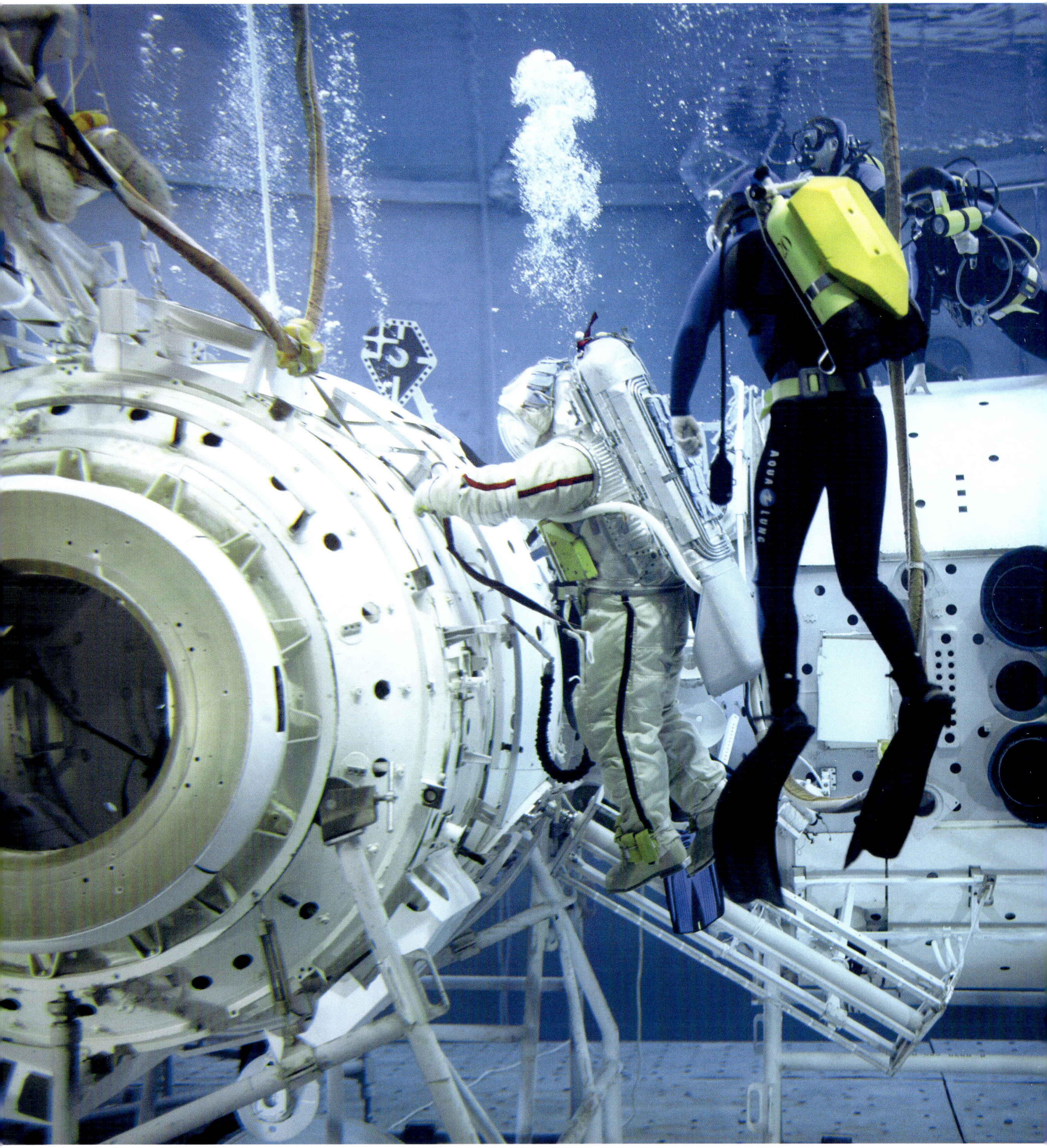

Cosmonaut, A N Shkaplerov, training session.

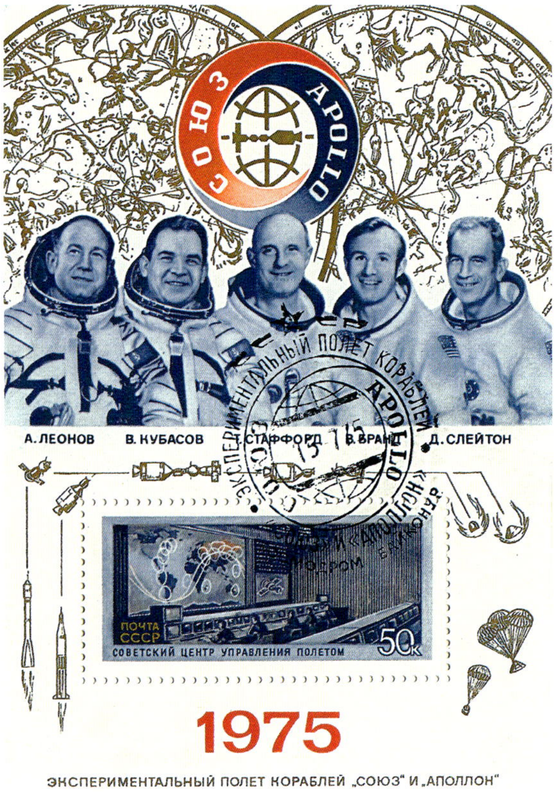

Postage stamp commemorating Apollo-Soyuz test project, July 1975.

Everyone of us, of course, dreamt of being on board the Soyuz spacecraft, which was supposed to dock with Apollo. But it was completely unexpected, when I was offered the post of Commander of the crew. Once I got over the shock, I said: 'It's great, I am honoured, but I don't know a single word in English'.

A A Leonov

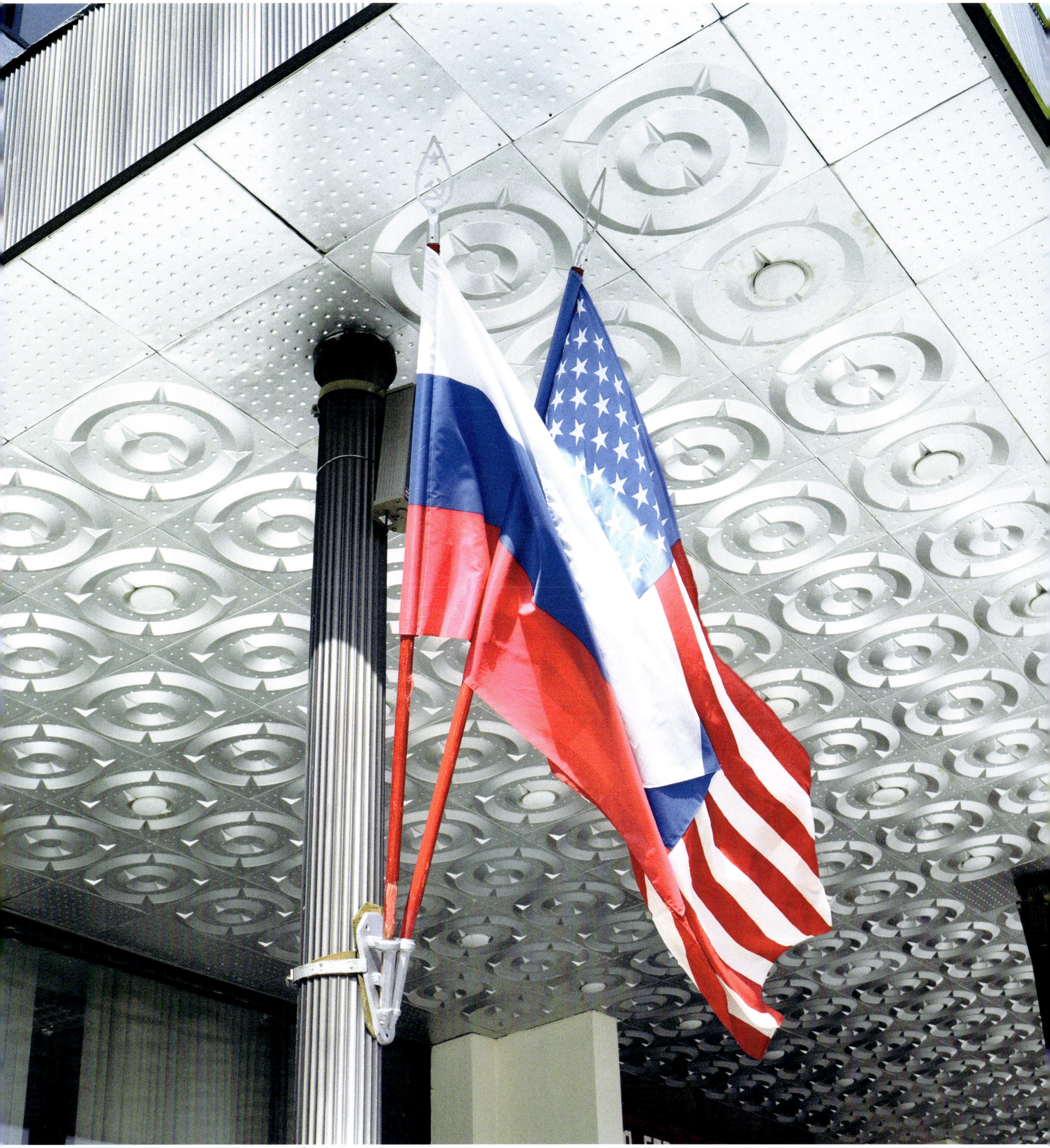

Flags, entrance to the House of Cosmonauts to greet Soyuz TMA–13 space flight crew, Star City.

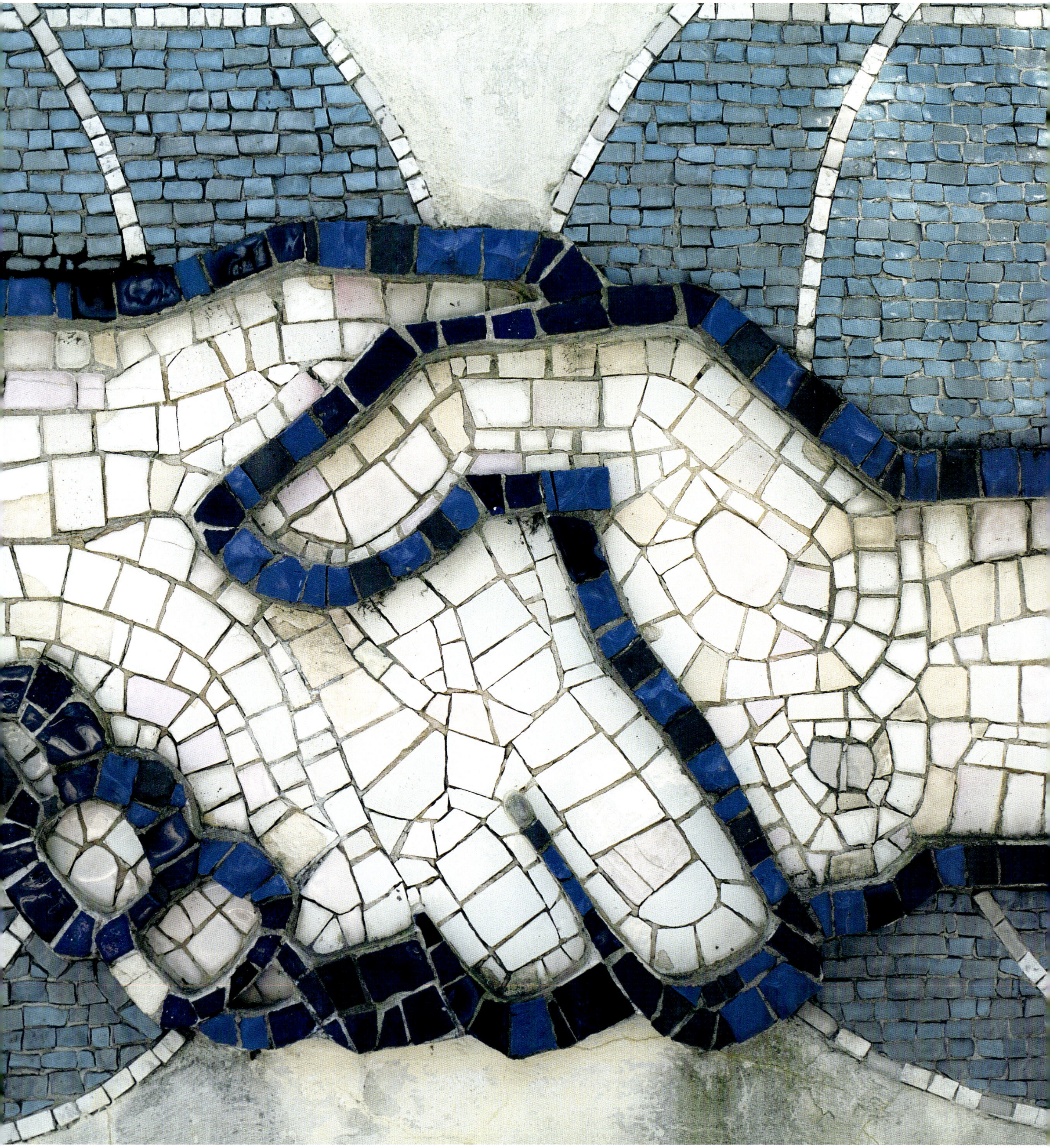

Decorative mosaic, detail, Star City.

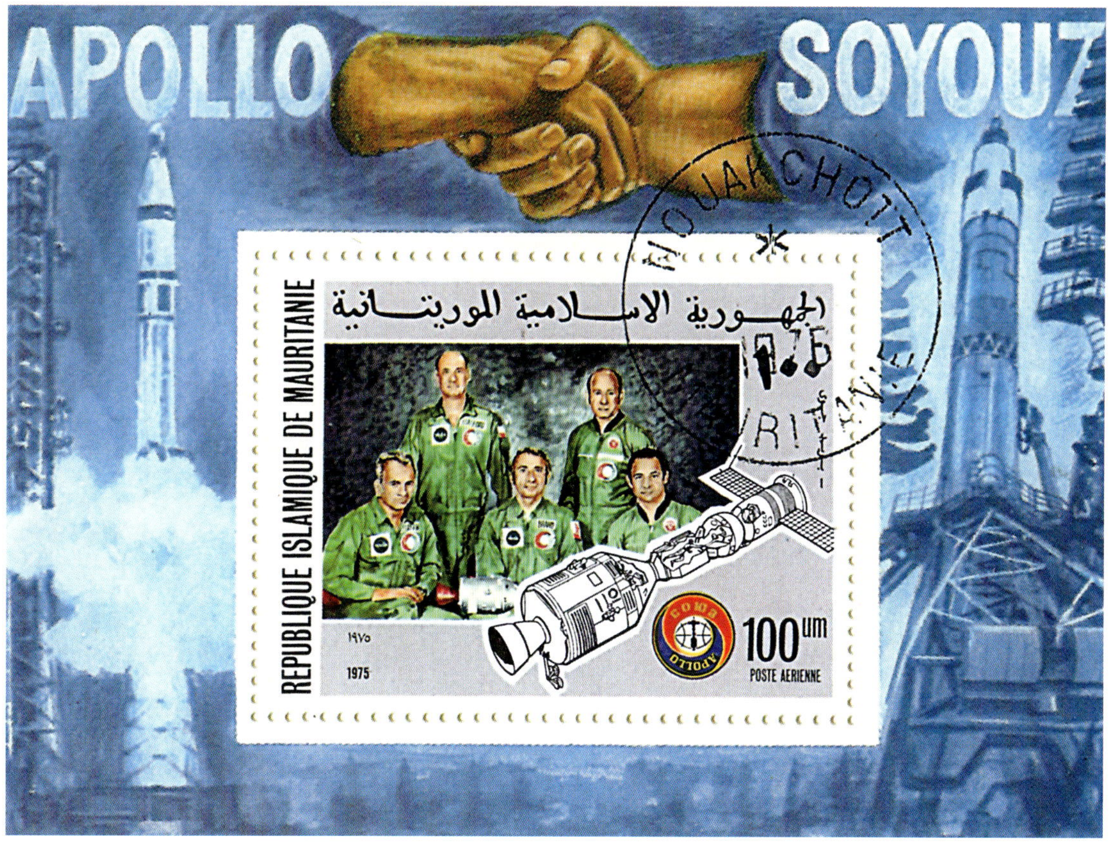

Above: Postage stamp from the Islamic Republic of Mauritania commemorating Apollo-Soyuz test project, July 1975.

Below right: Cosmonaut, A A Leonov with his wife and daughter.

Below left: Polish postage stamp featuring Leonov.

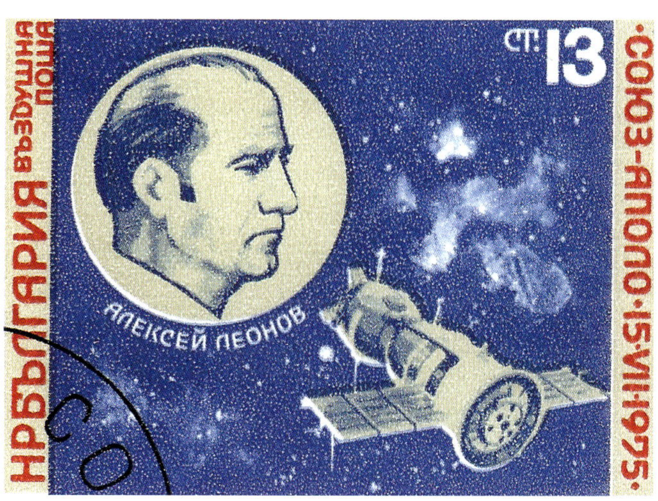

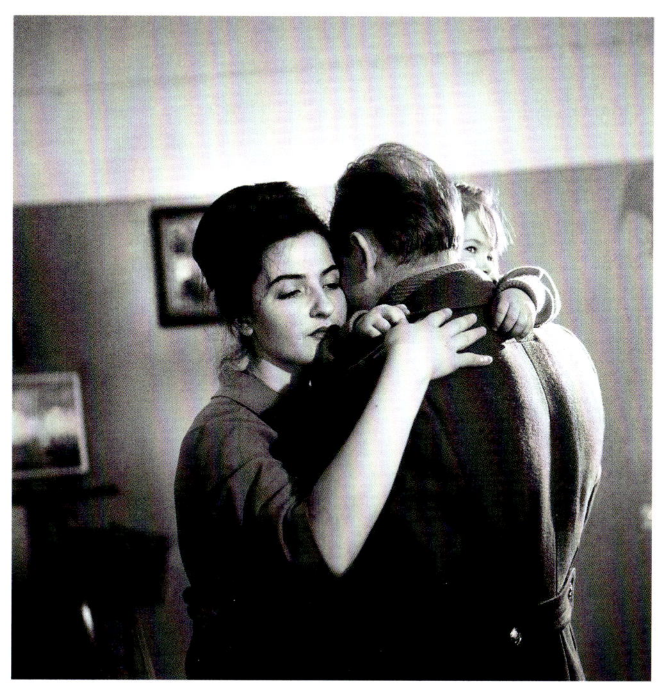

Below: Postage stamp, 1976.

Bottom: Postage stamp commemorating the launch of Zond–5 on 15 September 1968.

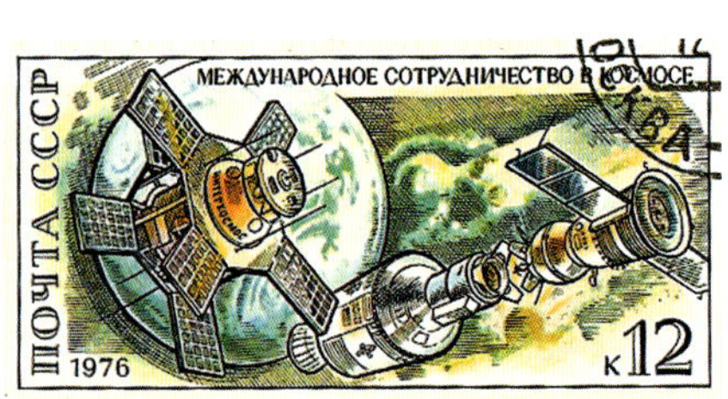

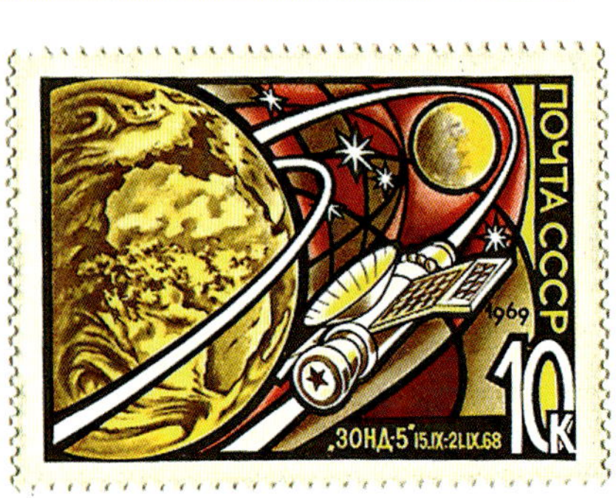

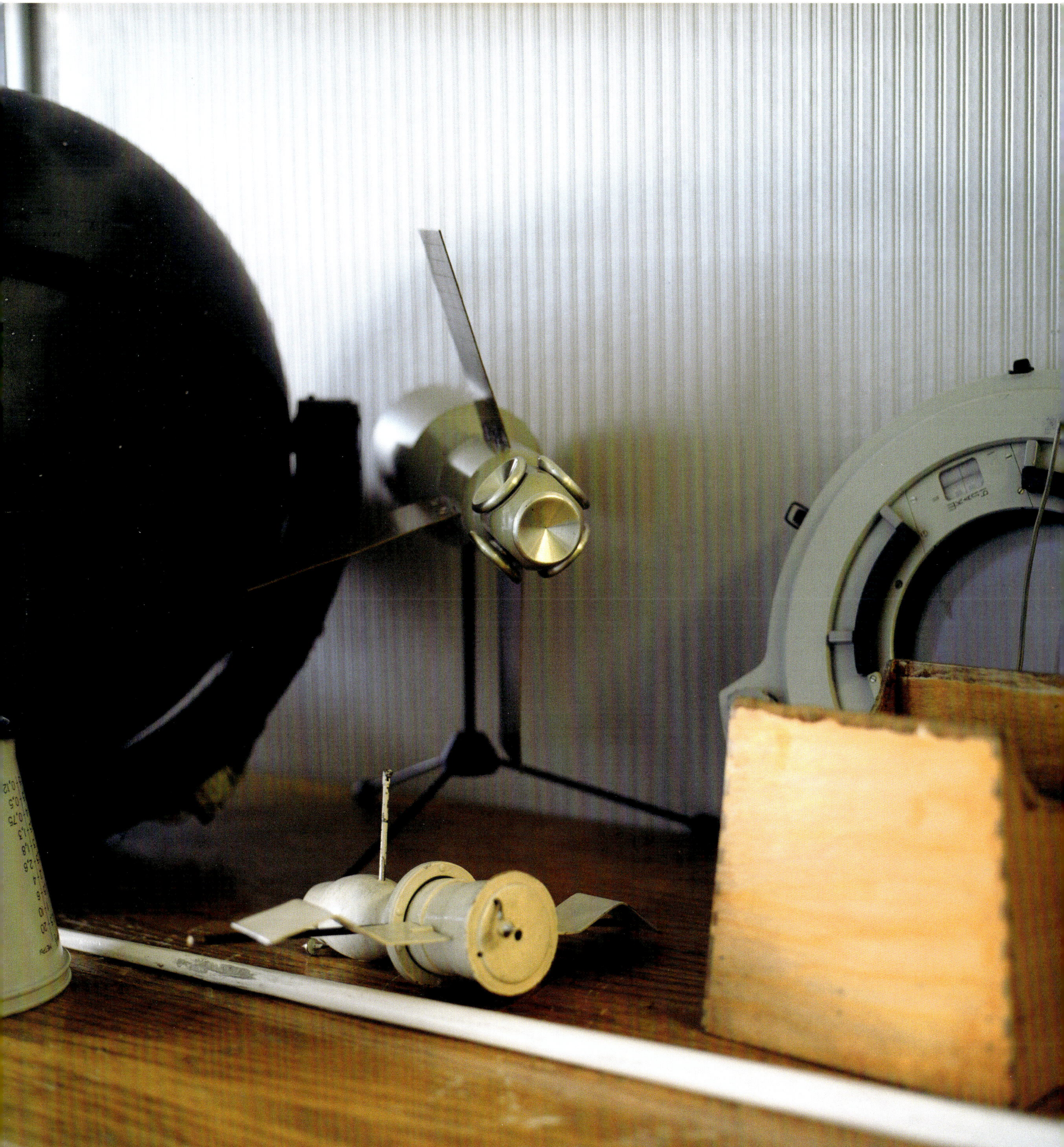

Mock-ups, Planetarium.

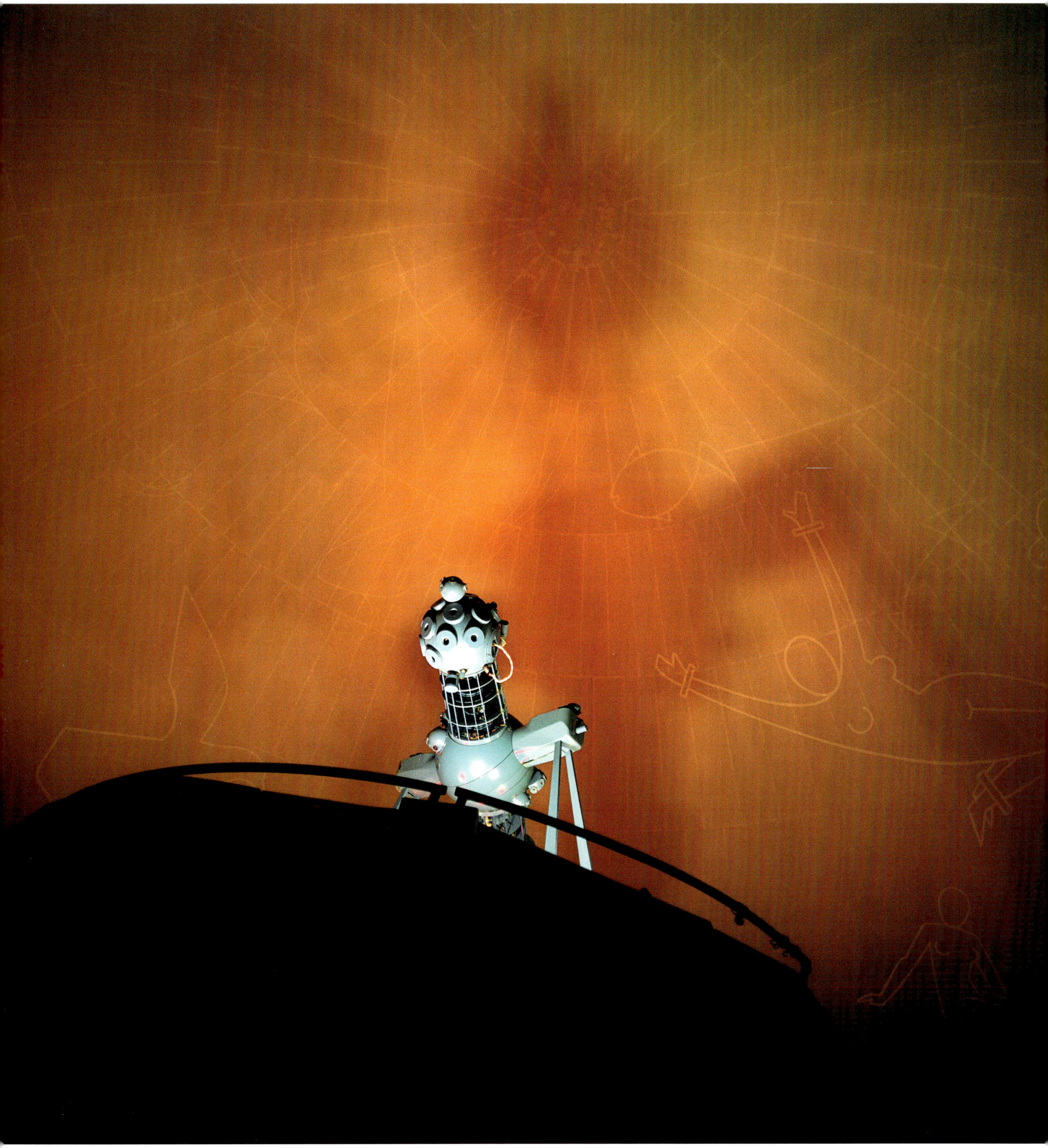

Planetarium.

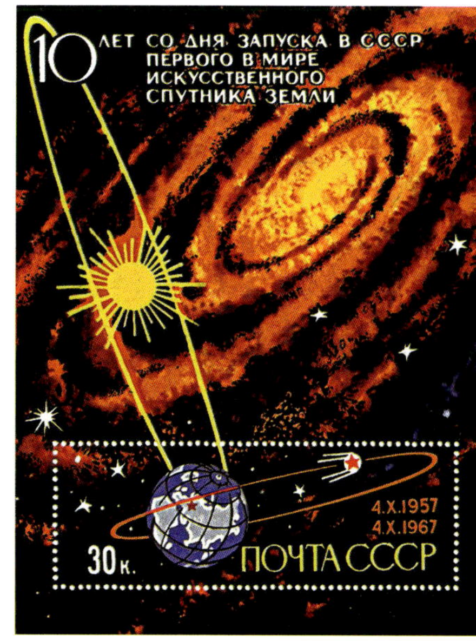

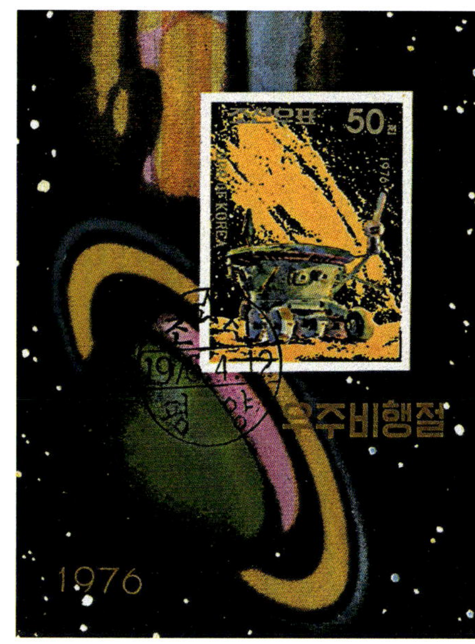

Top: Postage stamp commemorating the 10th anniversary of Satellite–1 launch.

Bottom: Korean postage stamp, 1976.

Museum installation, House of Cosmonauts, Star City.

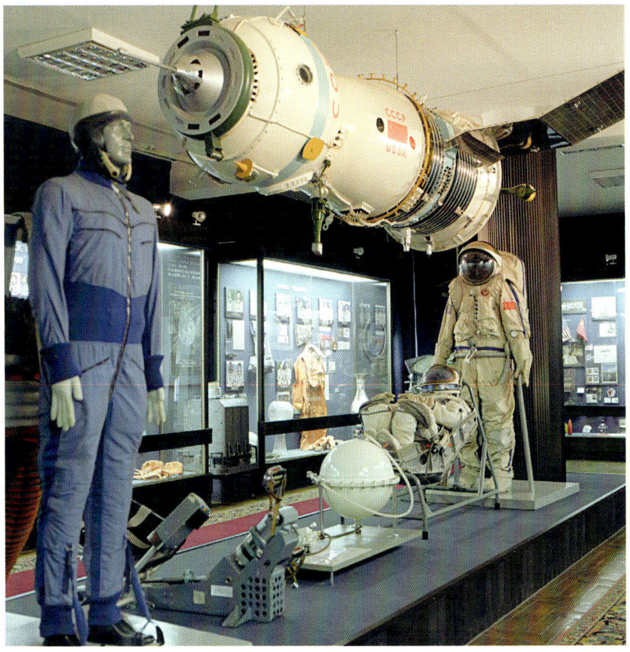

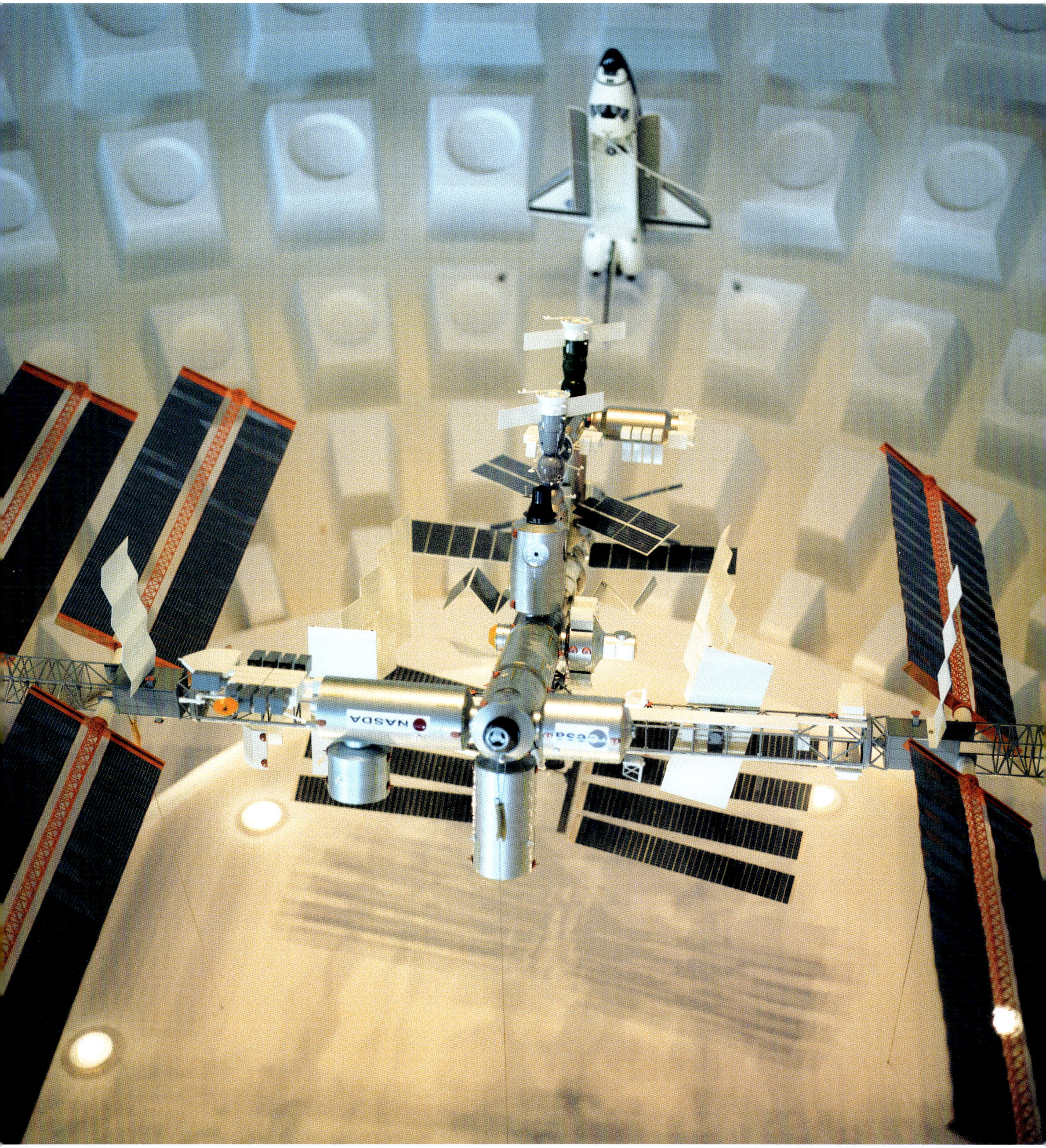

International Space Station (ISS) mock-up, Baikonur city museum.

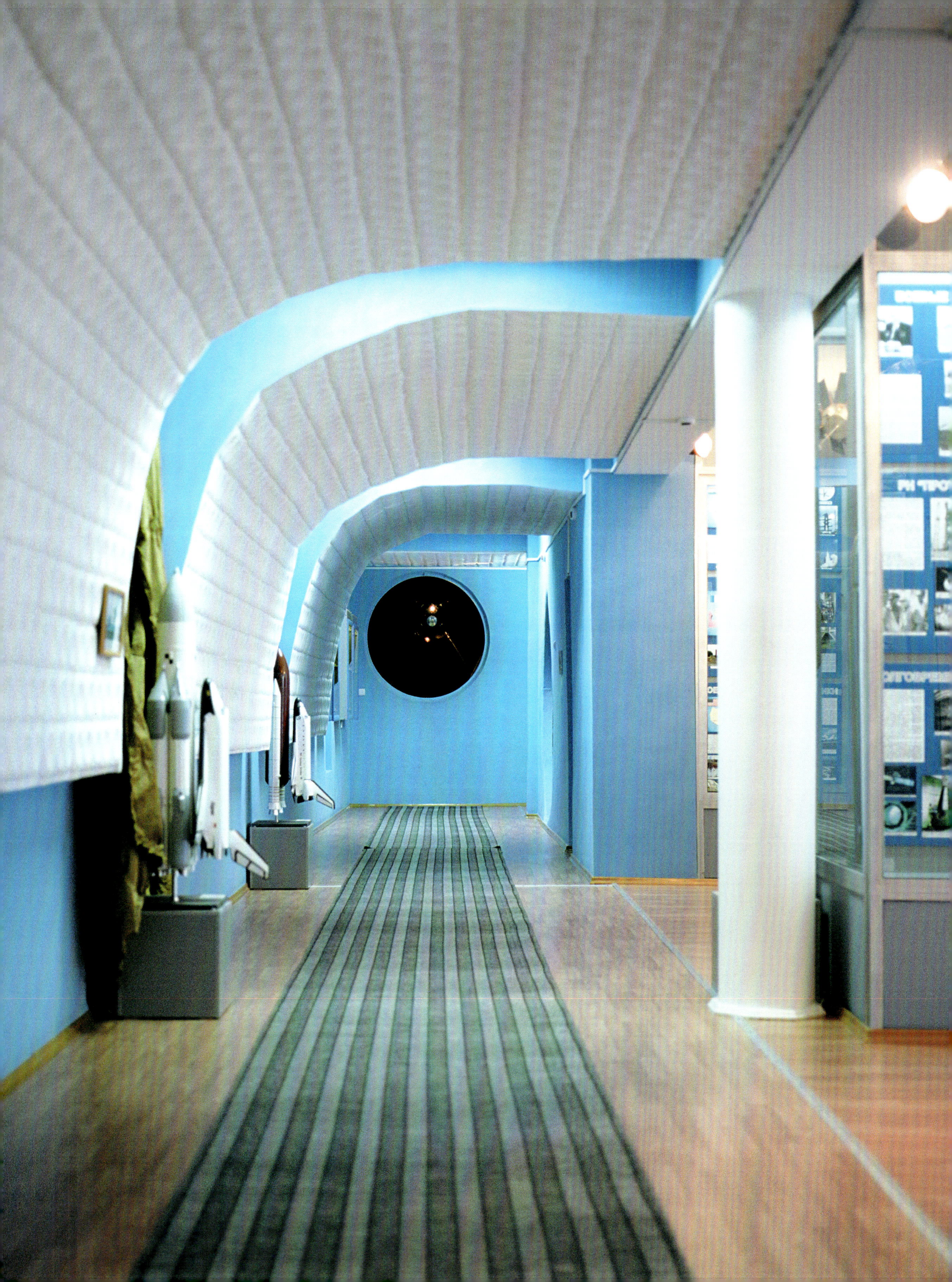

Hallway, Baikonur city museum.

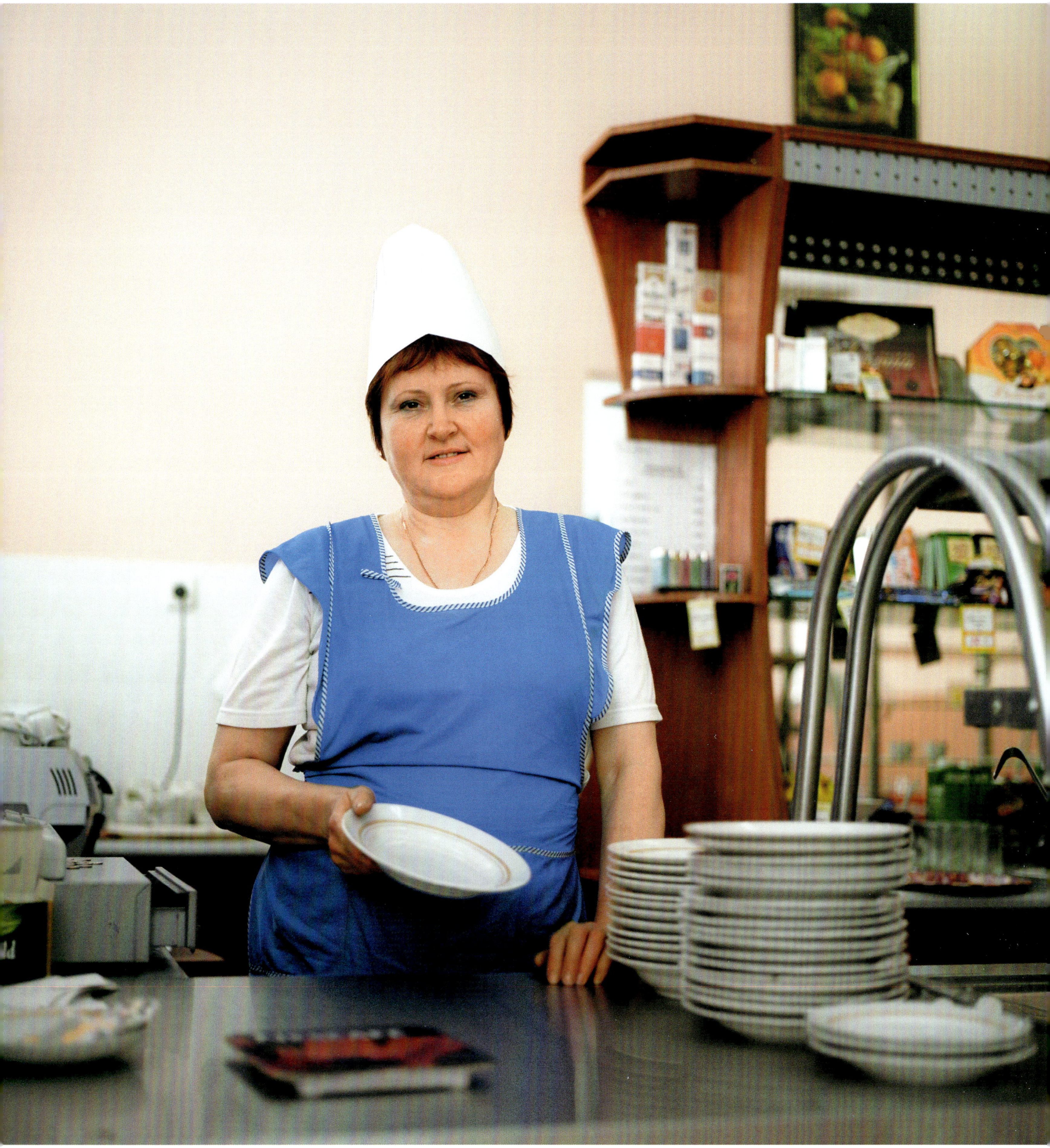

Cook, N Fedorova.

Canteen.

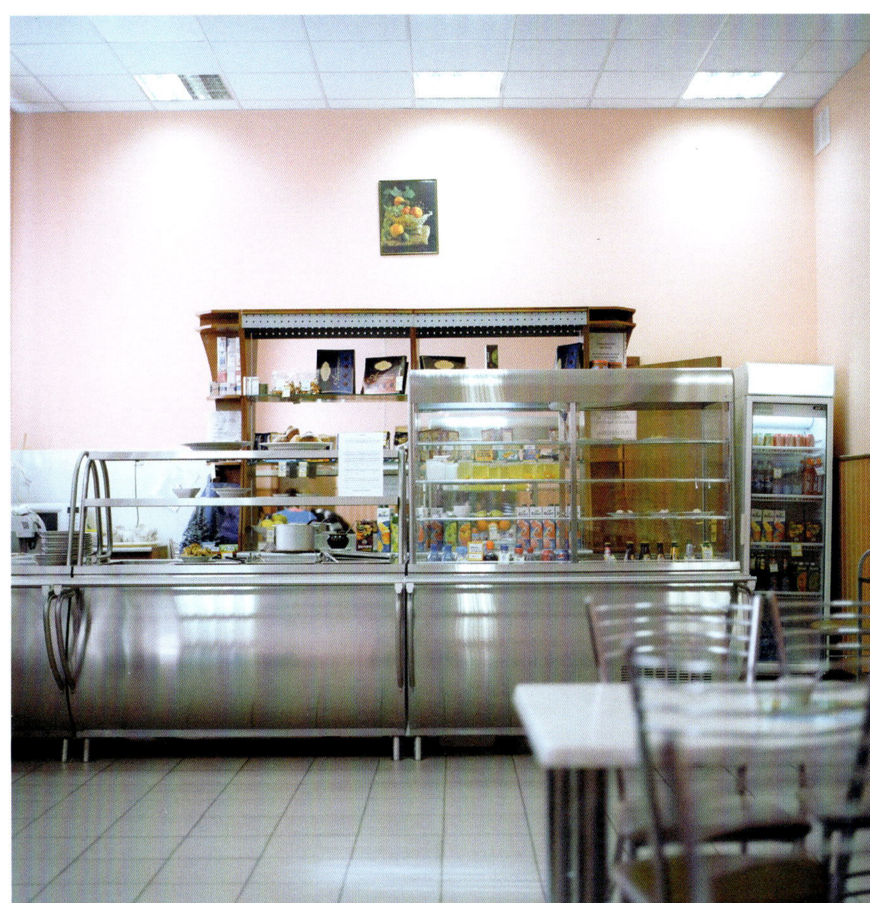

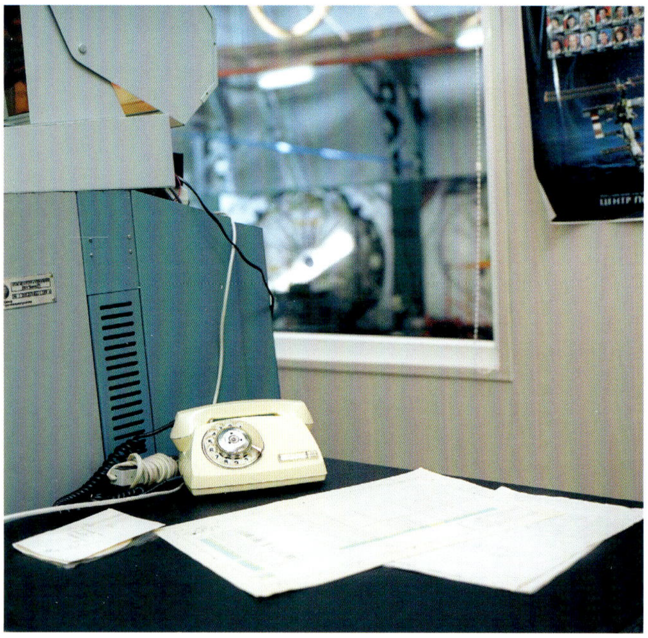

Left and below: Control room, Vykhod specialised simulator used for training cosmonauts to manage EVA-suit systems, controlling its independent and on-board systems for life-support and air-locking in standard operational use and in off-nominal situations.

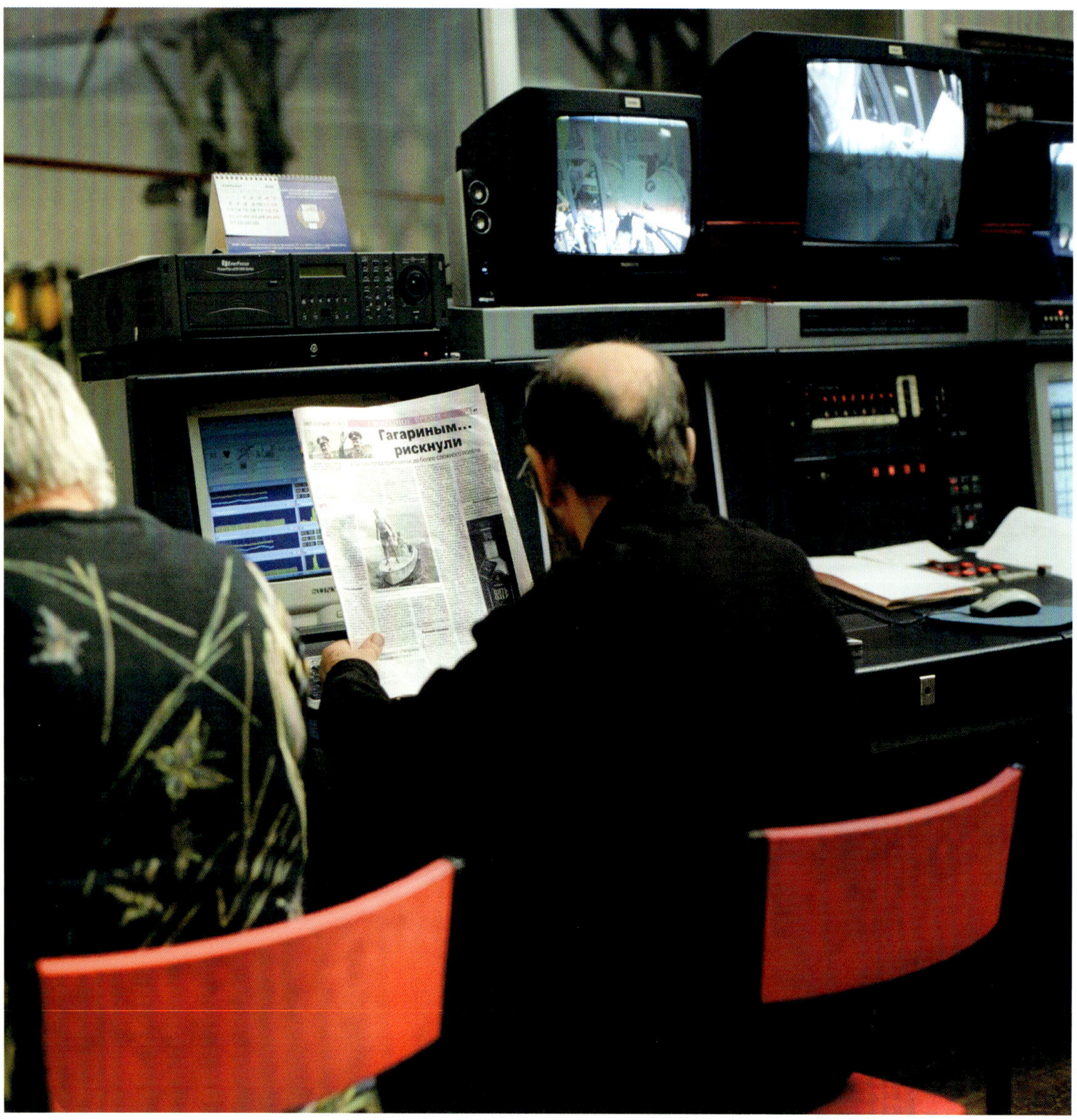

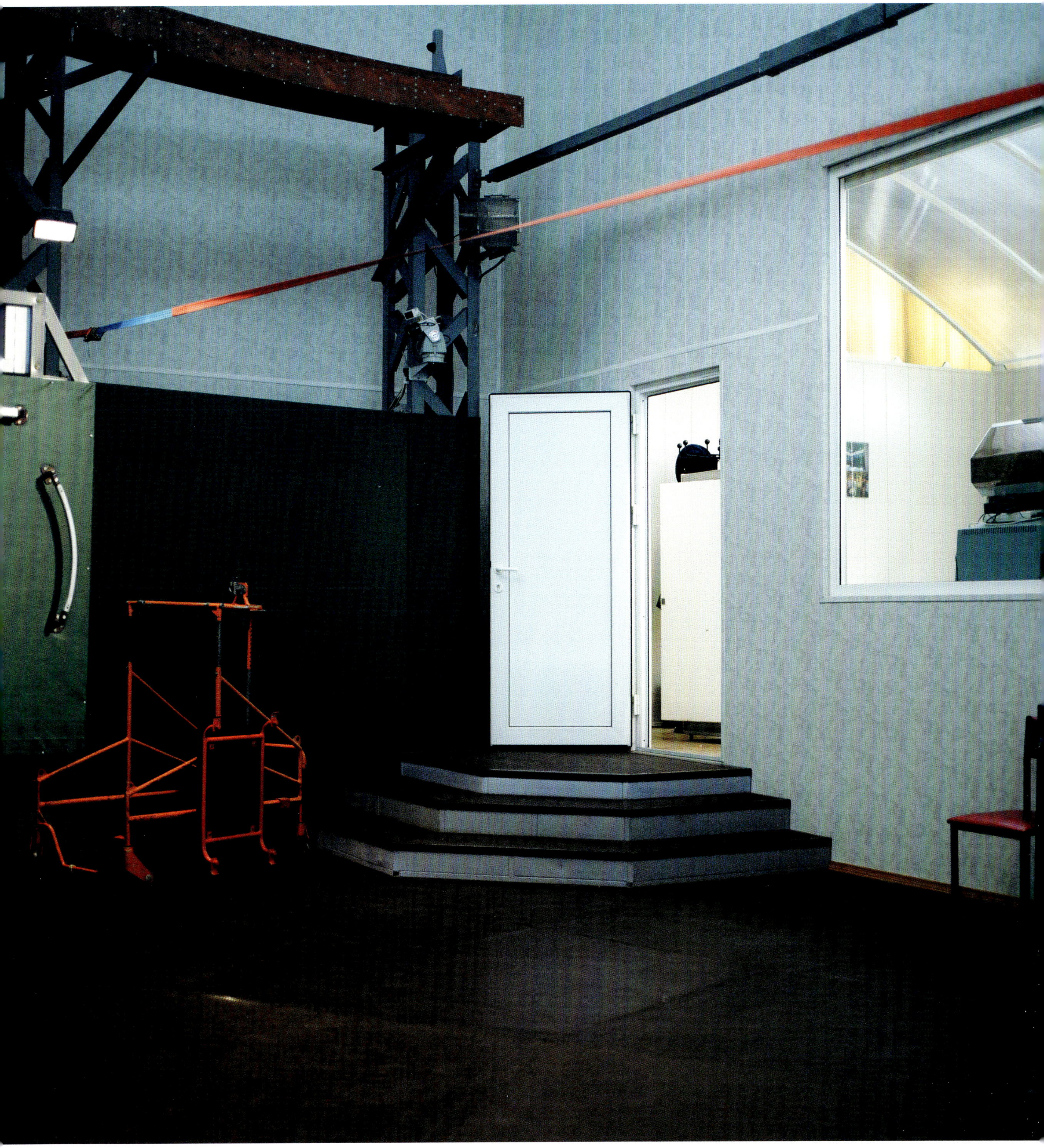

Entrance to control room, Vykhod specialised simulator.

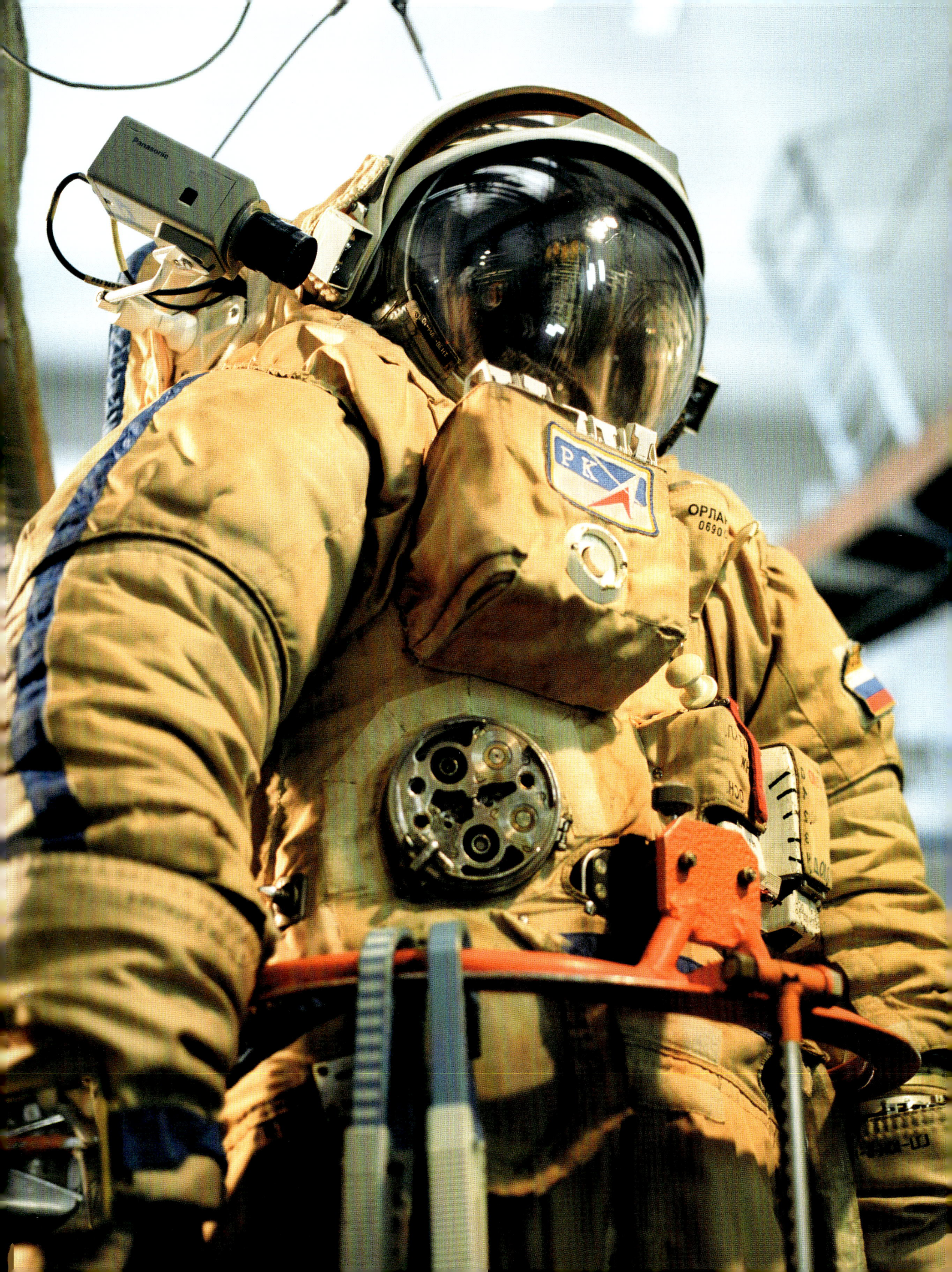
Panasonic
ОРЛАН

Left: Space suit used in Vykhod simulator.
Below: Vykhod simulator, control panel.

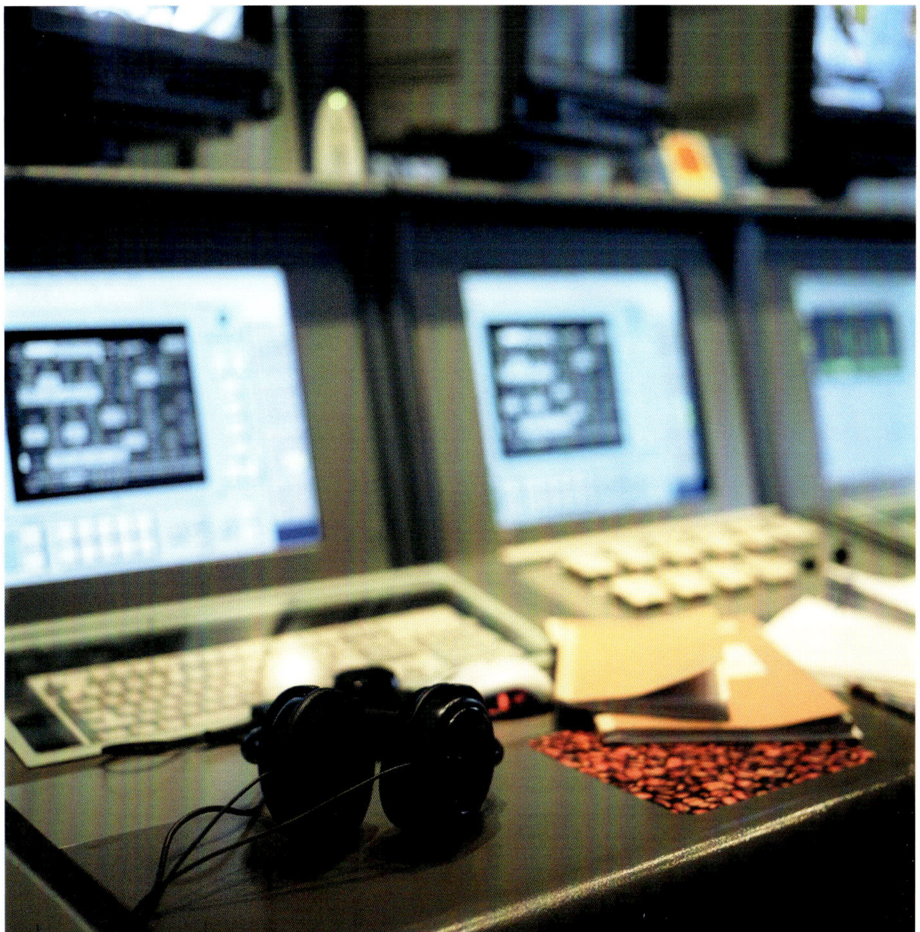

We are going into space, because without humanity it will always be alien and distant. We have to make it more earthly...

G M Grechko

Cosmonauts, G M Grechko and Y V Romanenko, Soyuz –26 space flight crew, launched 16 January 1978.

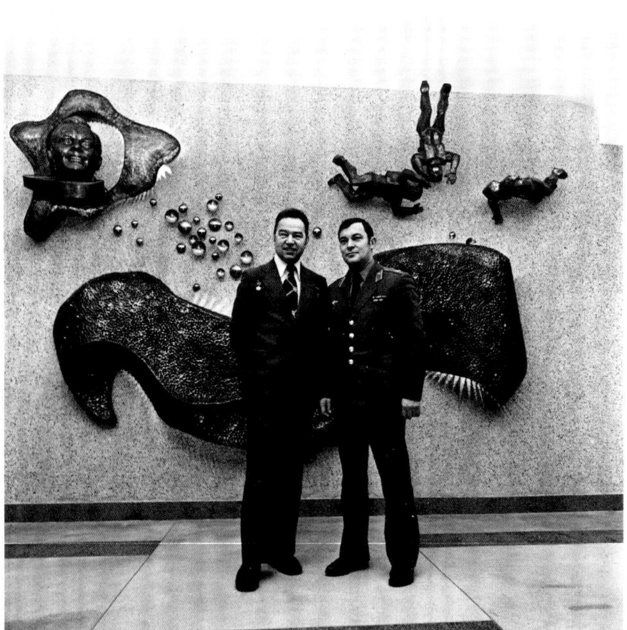

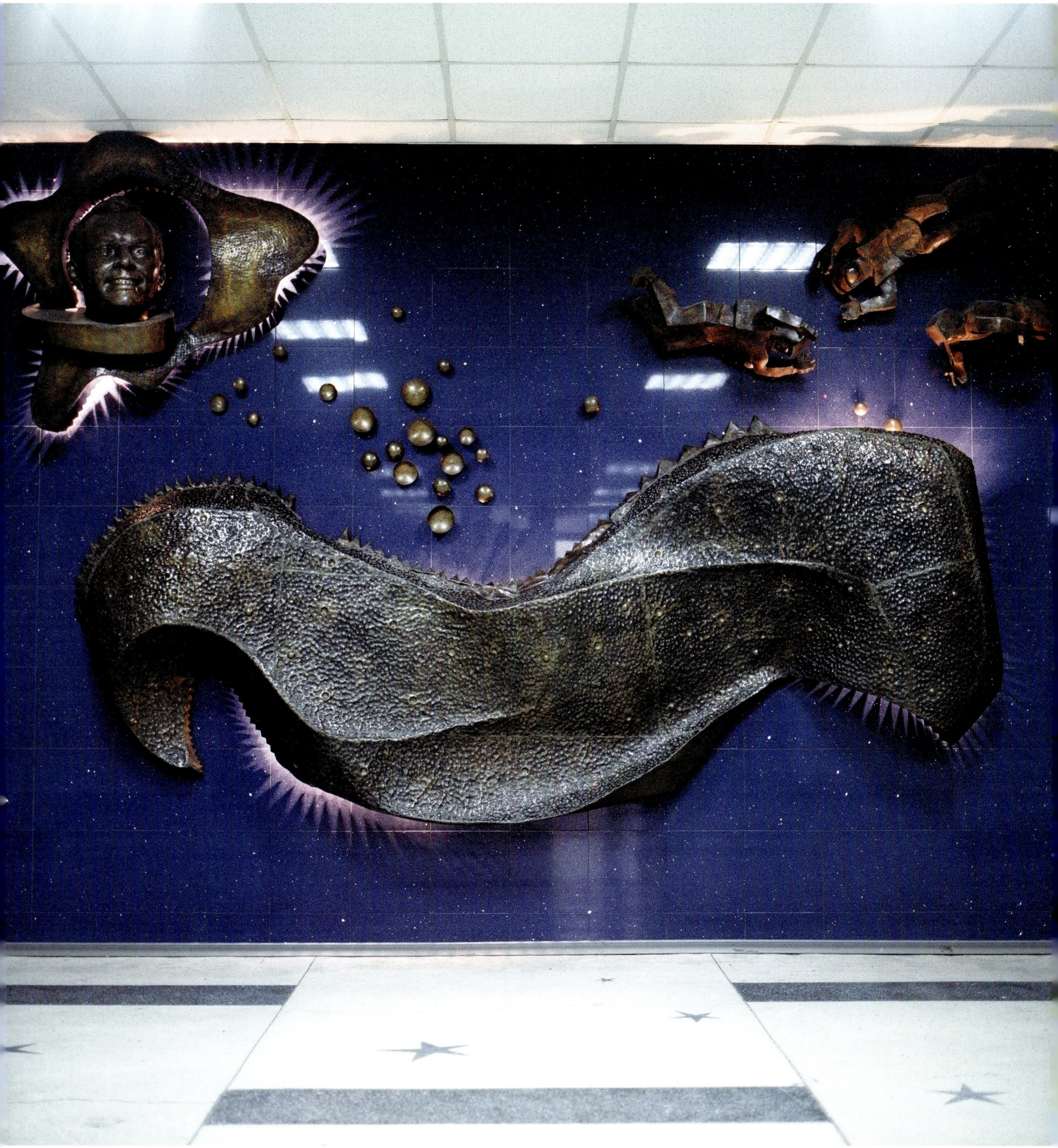

Main entrance hall, House of Cosmonauts, Star City.

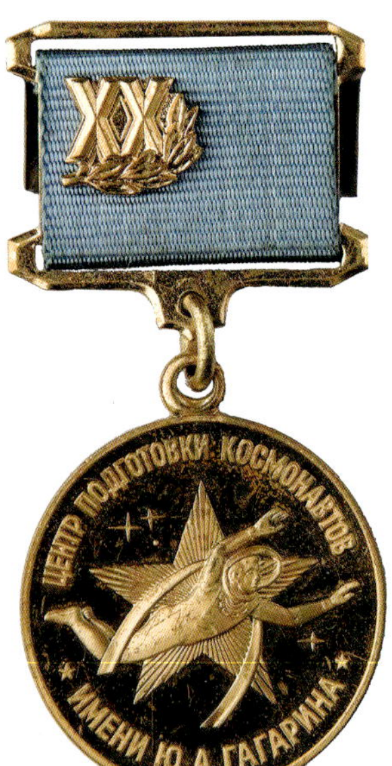

Left: Medal commemorating the 20th anniversary of the Yuri Gagarin Russian State Science Research Cosmonauts Training Centre.

Right: Sculpture of cosmonaut, at the entrance to the House of Cosmonauts.

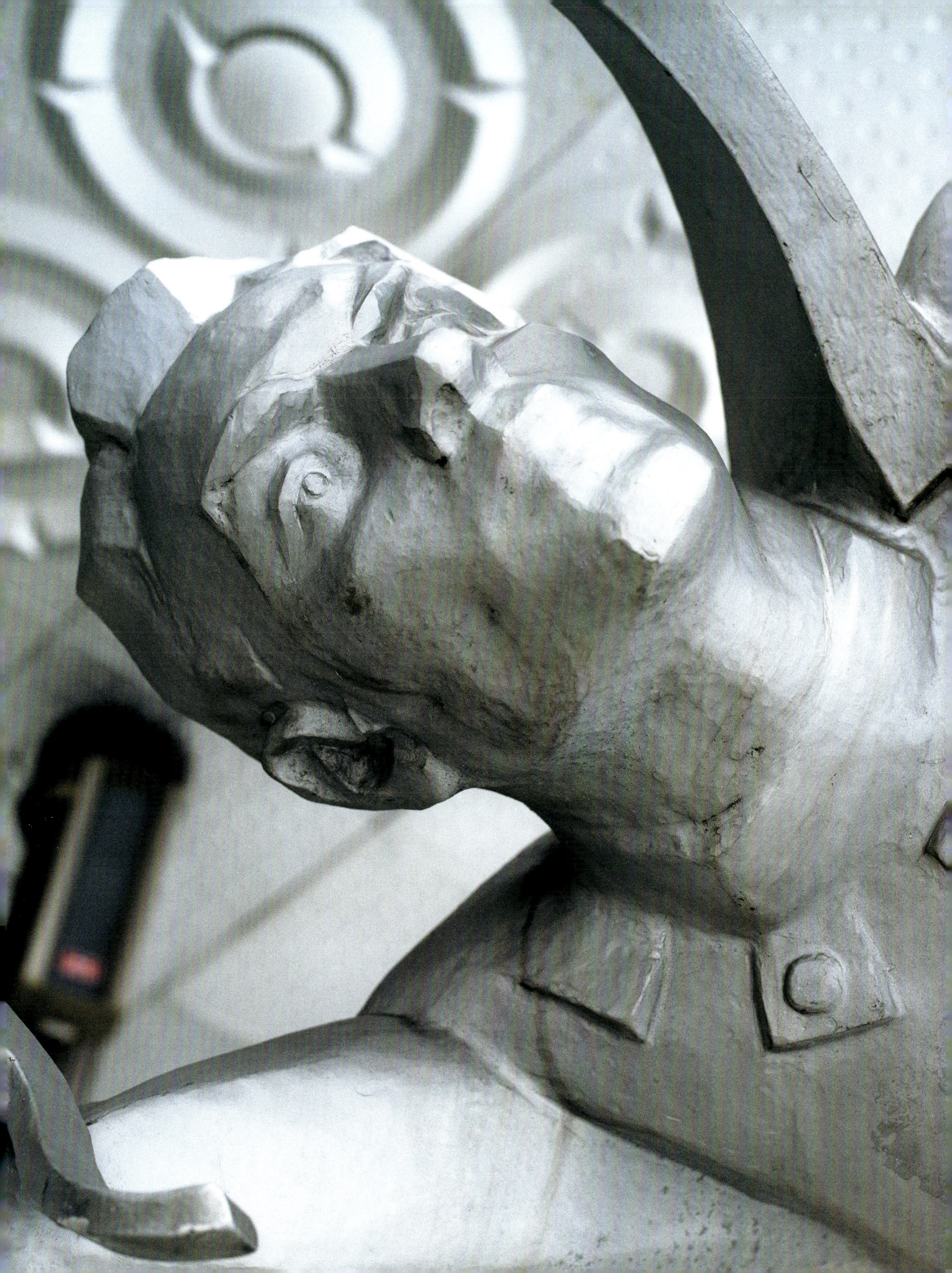

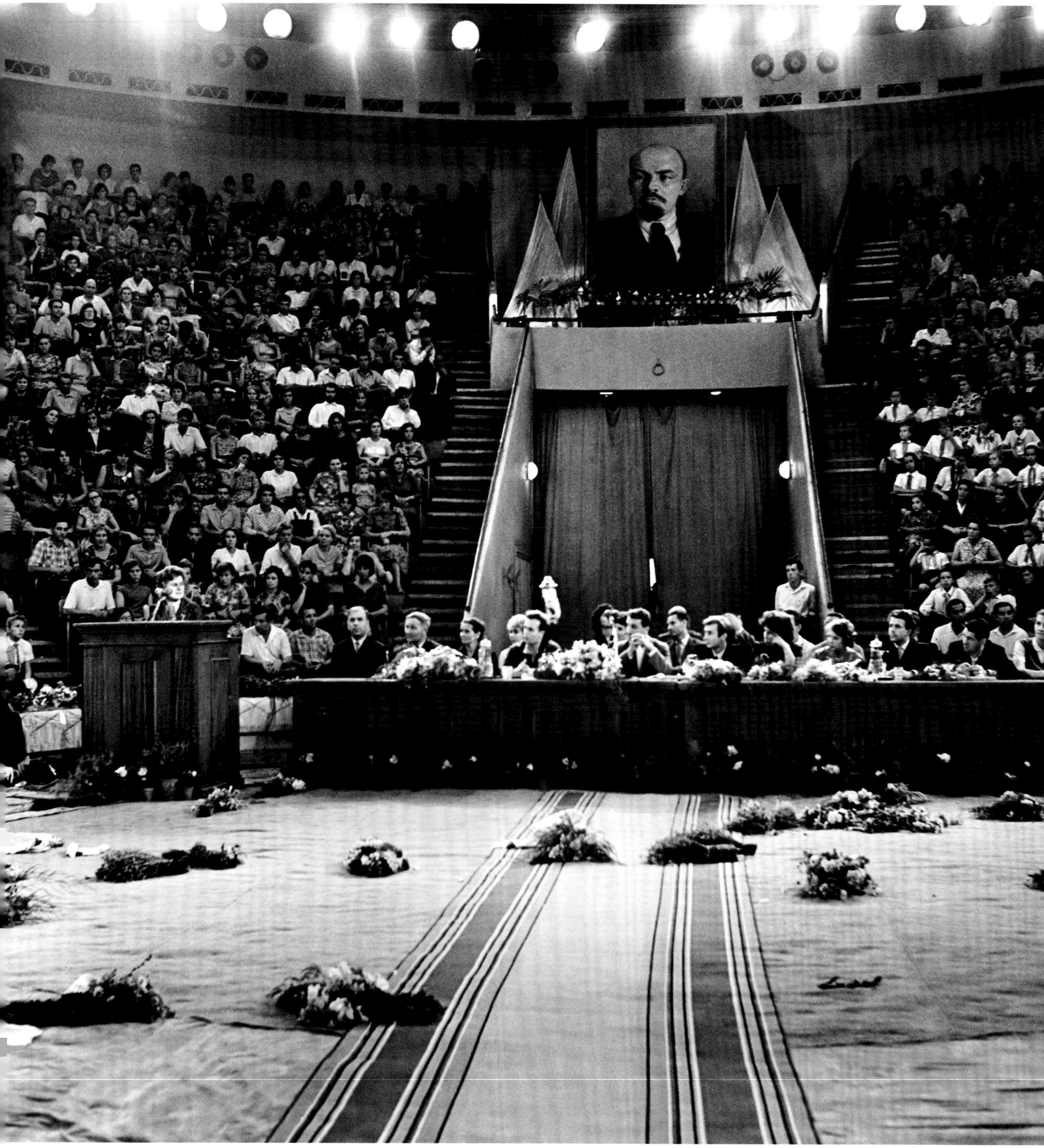

Official ceremony to welcome back V V Tereshkova, the first woman in space, Vostok-6 spacecraft, launched June 1963.

Right: Soyuz-16 space flight crew, launched 2 December 1974, with relatives and friends. The monument featuring Lenin was a gift from the Central Committee of the Komsomol. Before going on space missions, every crew member comes here to pose for a commemorative photograph.

Below: The official ceremony to welcome back the space flight crew of Soyuz TMA-13, Assembly Hall, House of Cosmonauts.

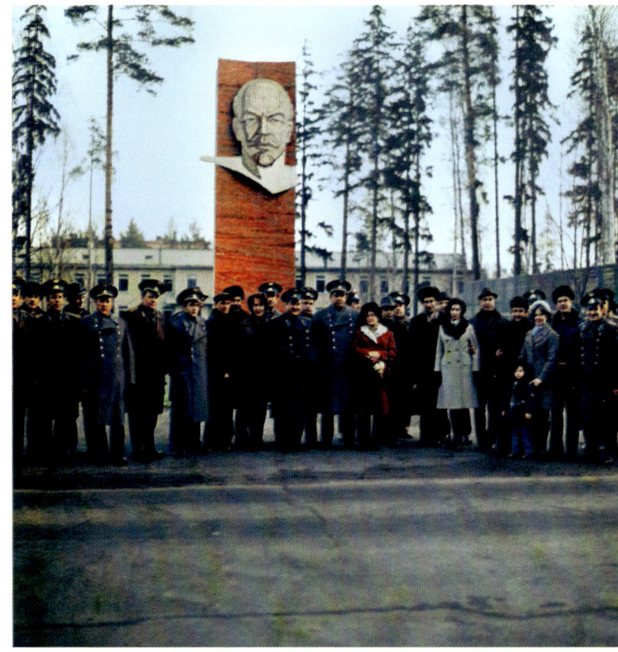

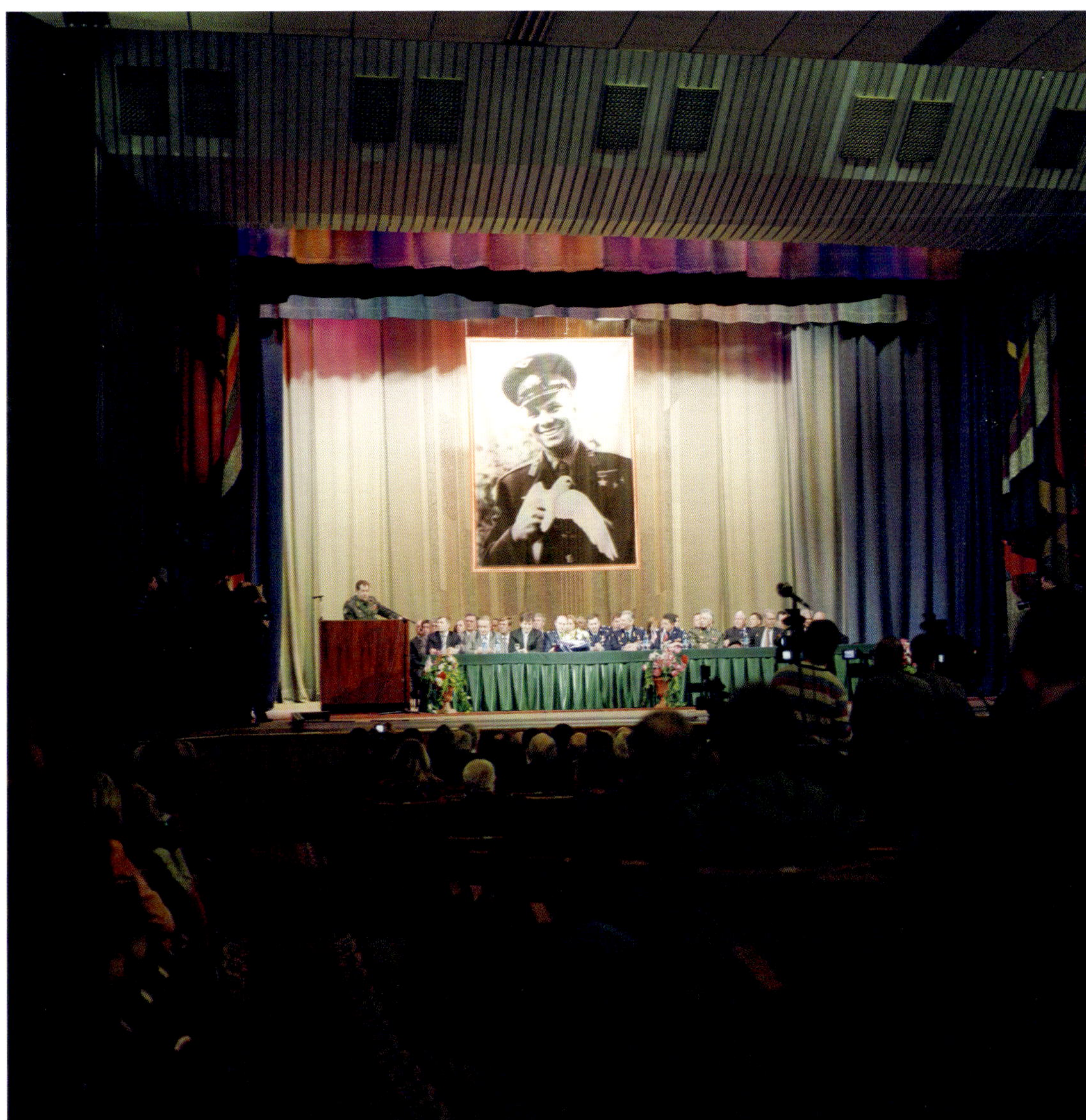

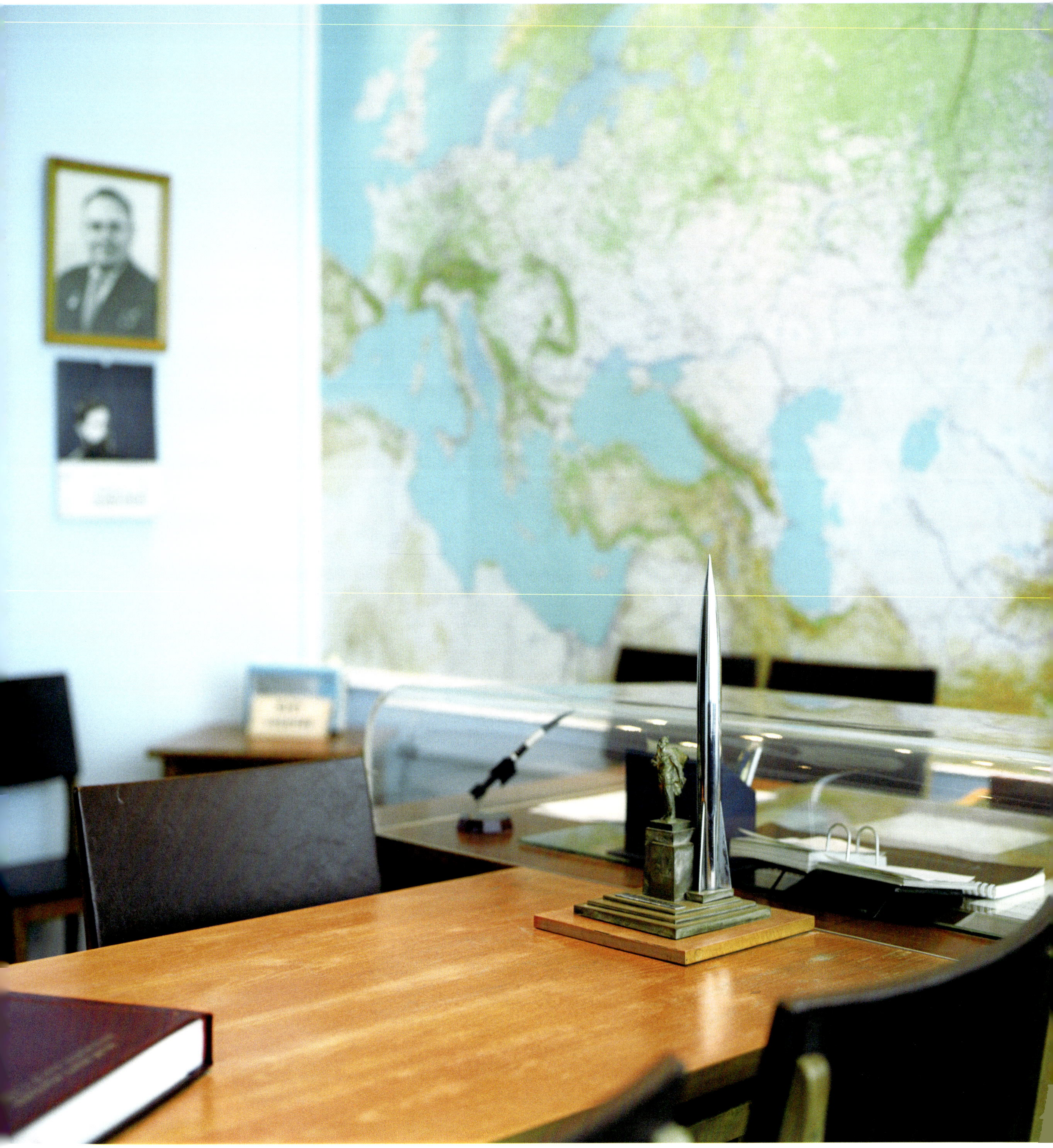

The office of Y A Gagarin (1934–1968), Museum of the House of Cosmonauts.

Cosmonauts, G M Grechko and Y V Romanenko, the space flight crew of Soyuz–26, launched 16 January 1978, in Gagarin's office at the House of Cosmonauts.

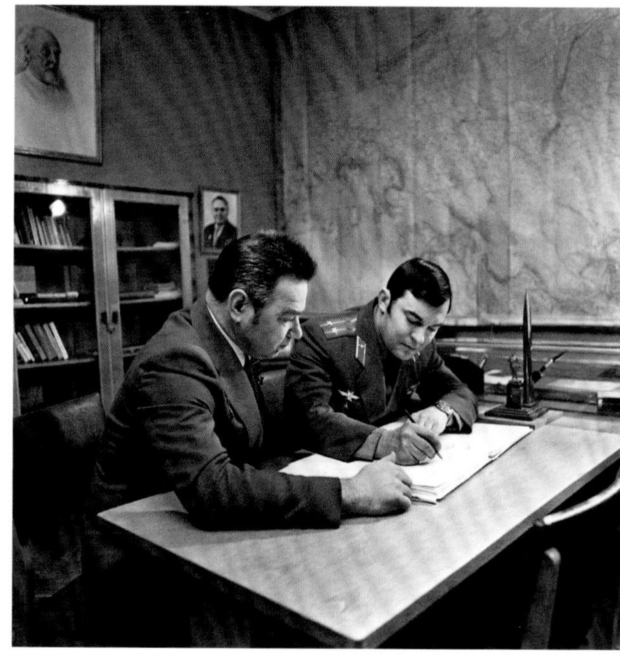

Since 1969, after completing preparation for another flight, each crew comes to the Museum of the House of Cosmonauts, to Gagarin's office to sign the commemoration album.

Office tables, Star City

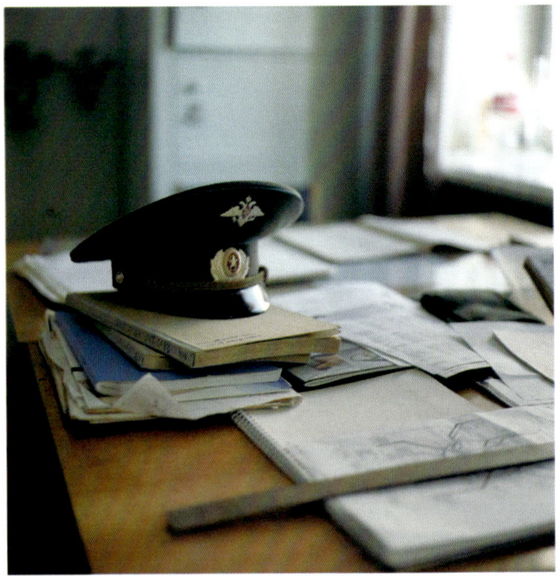

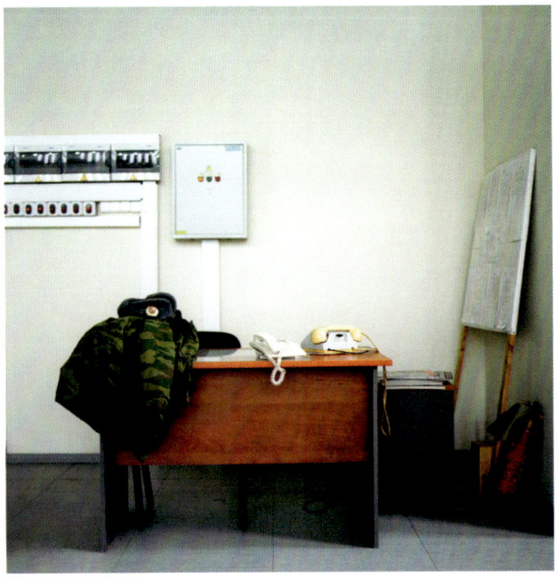

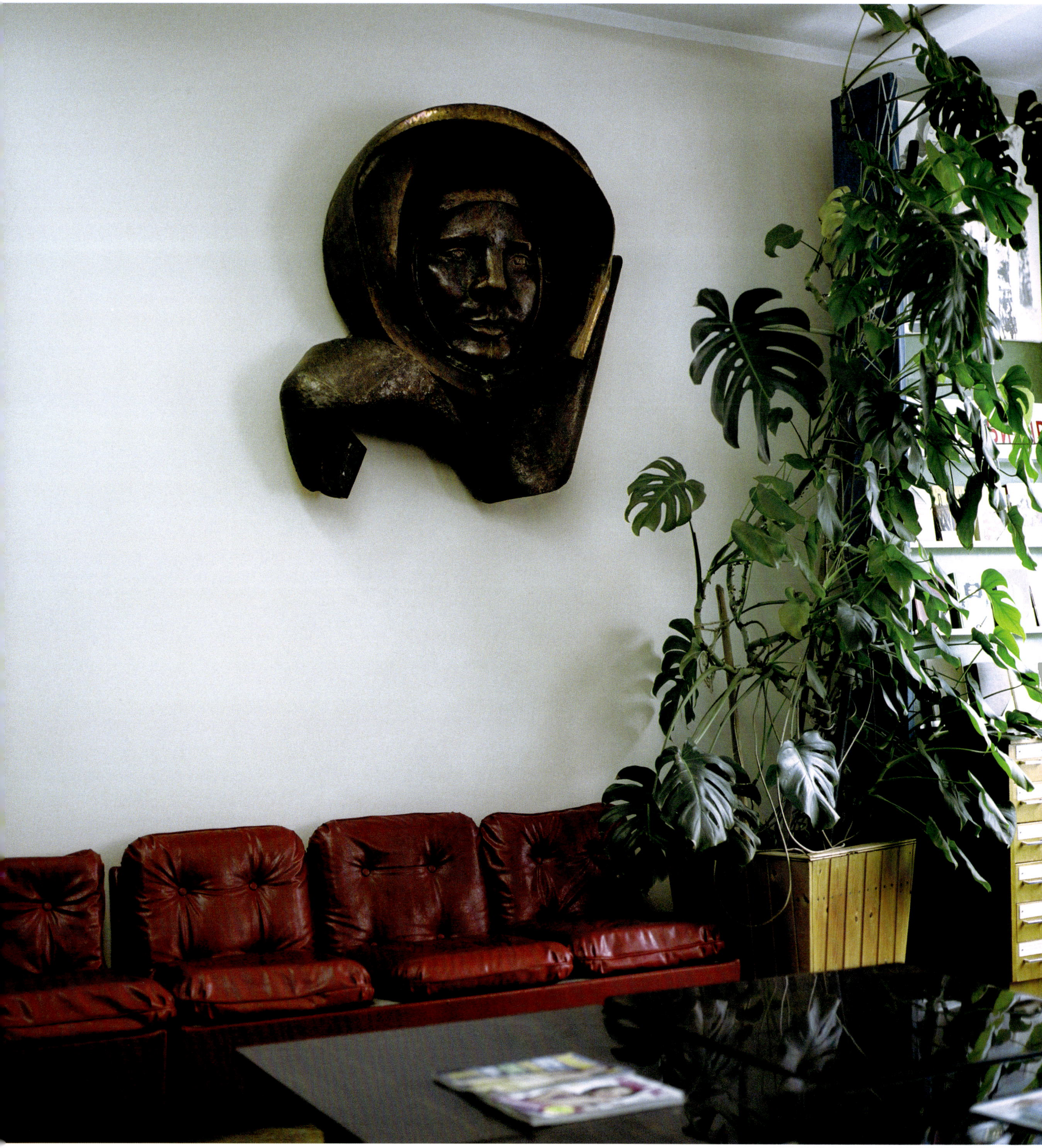

Library, House of Cosmonauts.

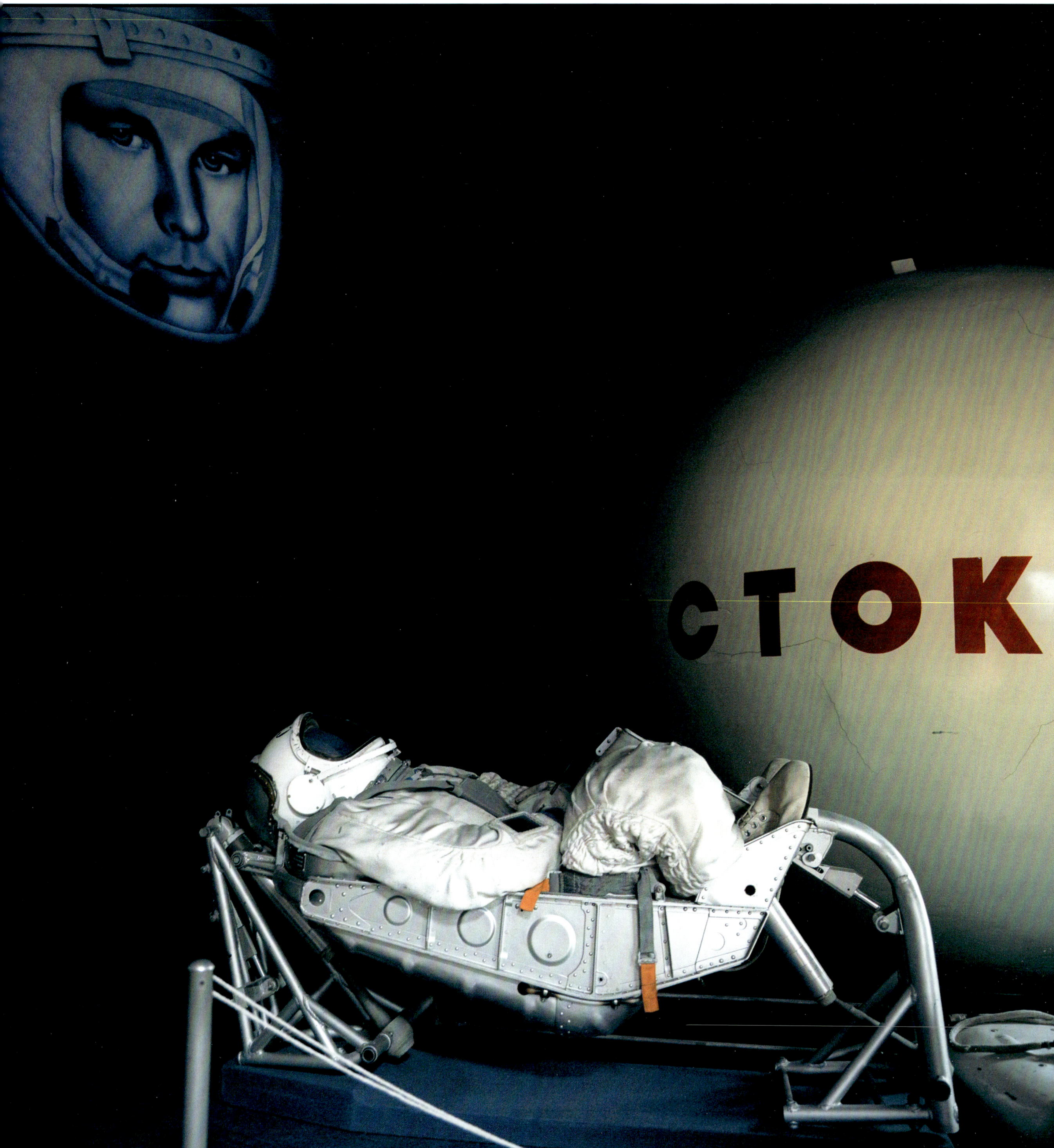

Vostok spacecraft, cosmonaut dummy, installation, Museum of the House of Cosmonauts.

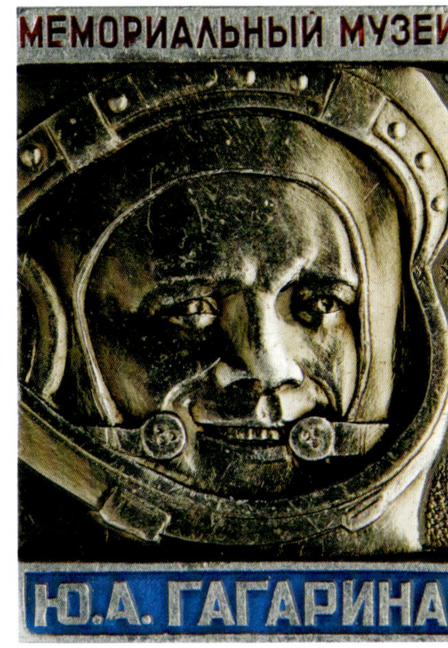

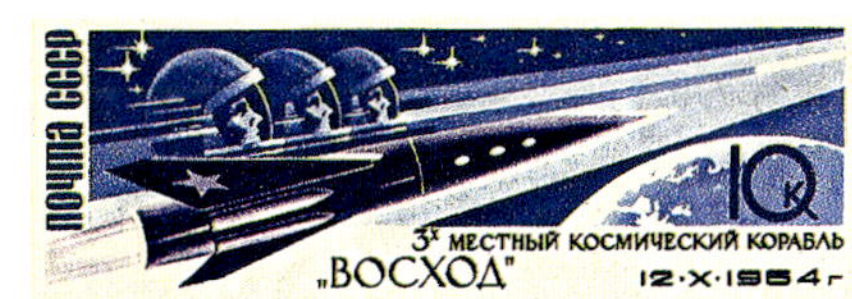

Top: Vintage badge from the Museum of Y A Gagarin.

Above: Postage stamp commemorating Voskhod–1, a spacecraft designed to accomodate three cosmonauts. This was the first space flight to carry more than one crewman into outer space, the first not to use spacesuits and to carry either an engineer or a physician into outer space.

Right: Decorative mosaic featuring Y A Gagarin, Star City.
Following page: Decorative mosaic, detail, Museum of the city of Baikonur.

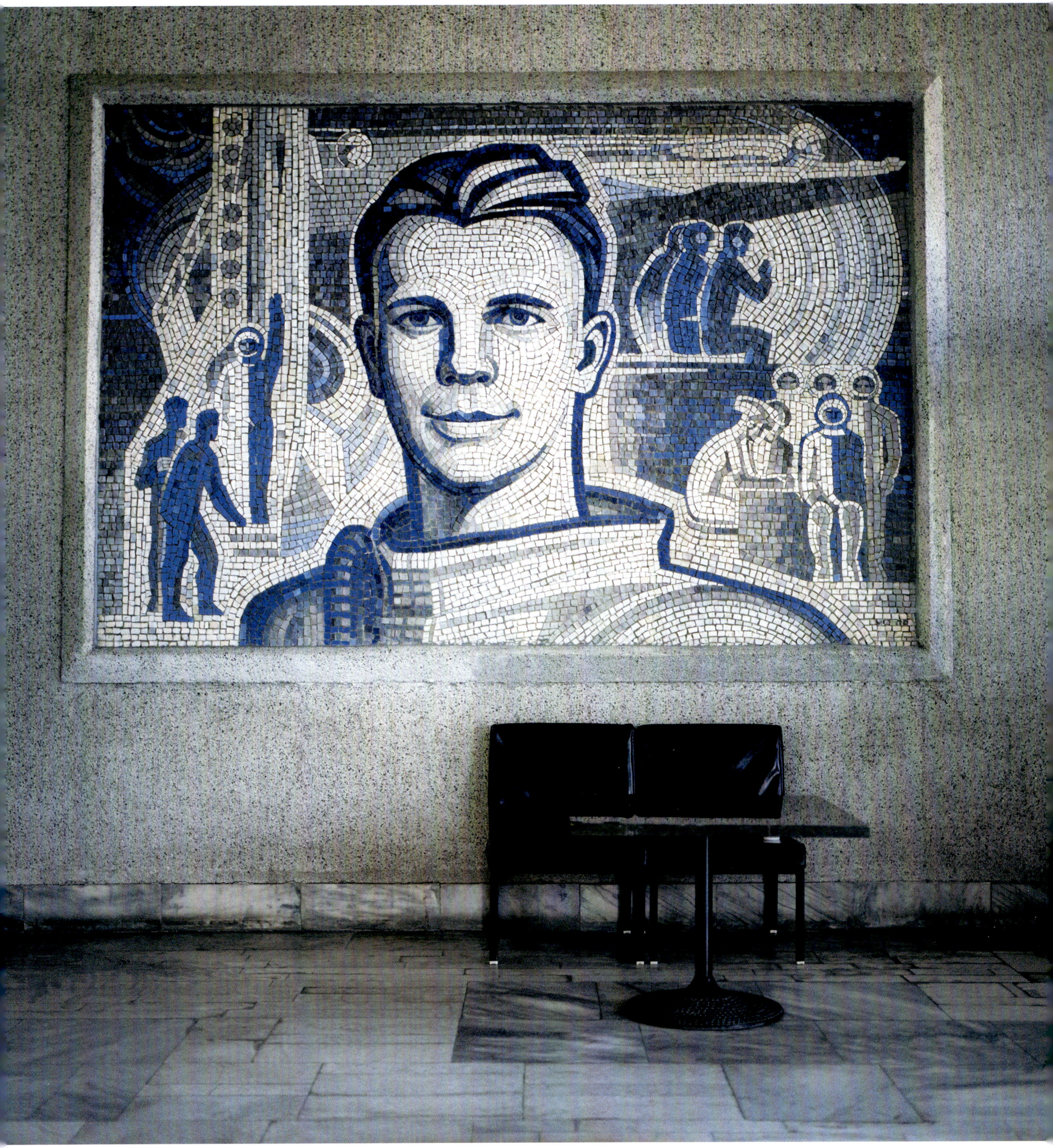

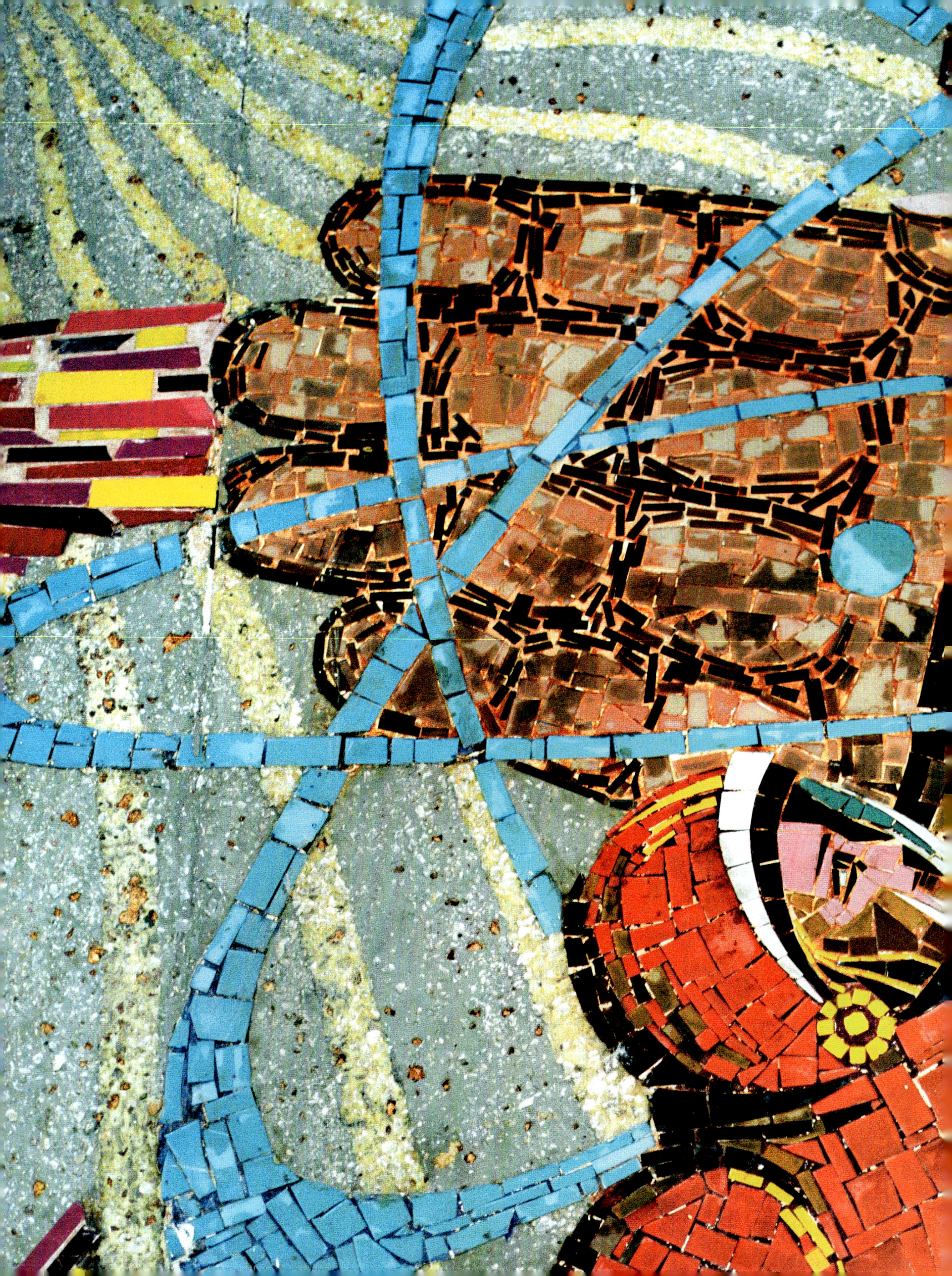

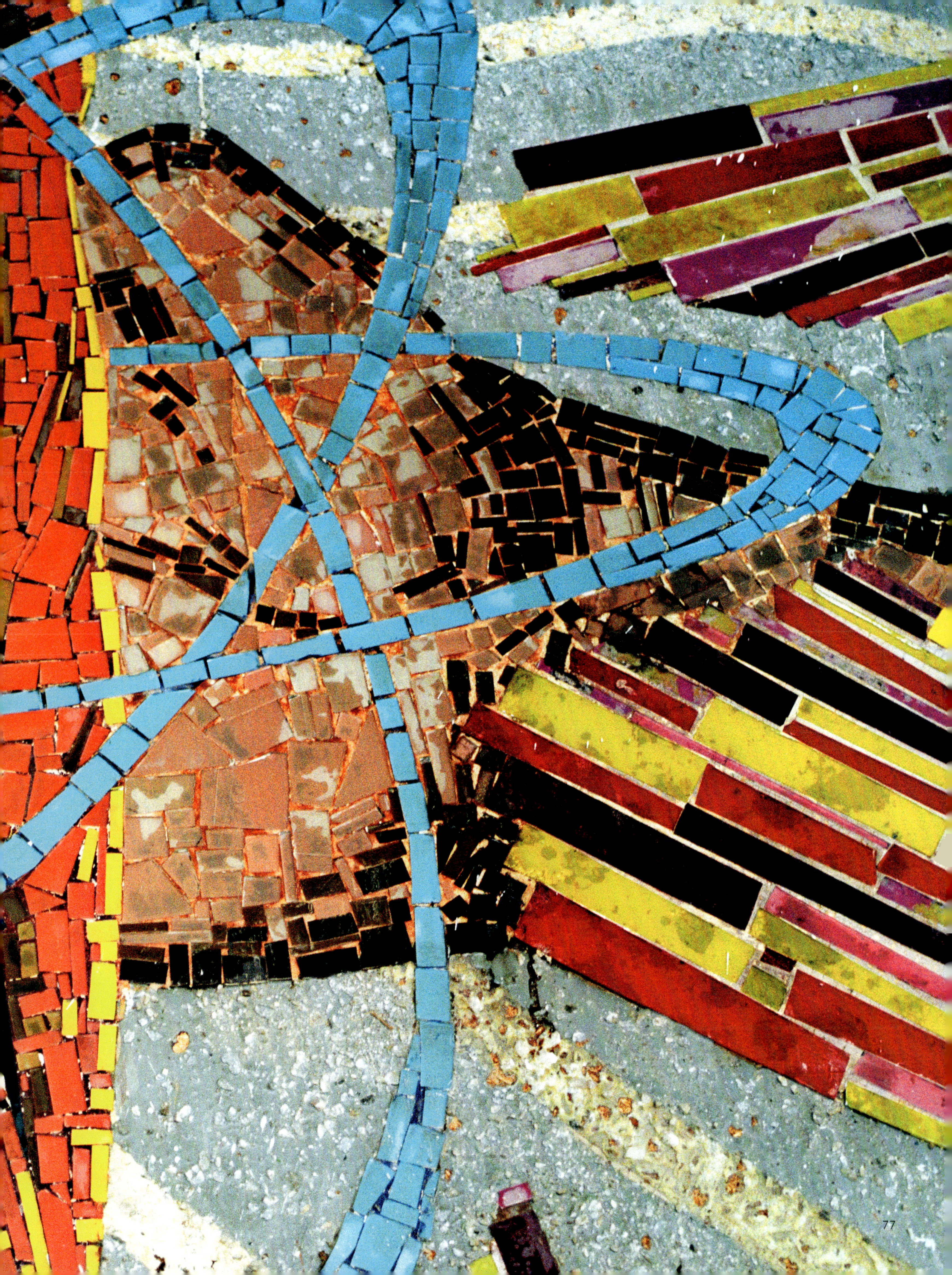

Obelisk Leninsk. From 1958–1995 Baikonur was known as Leninsk.

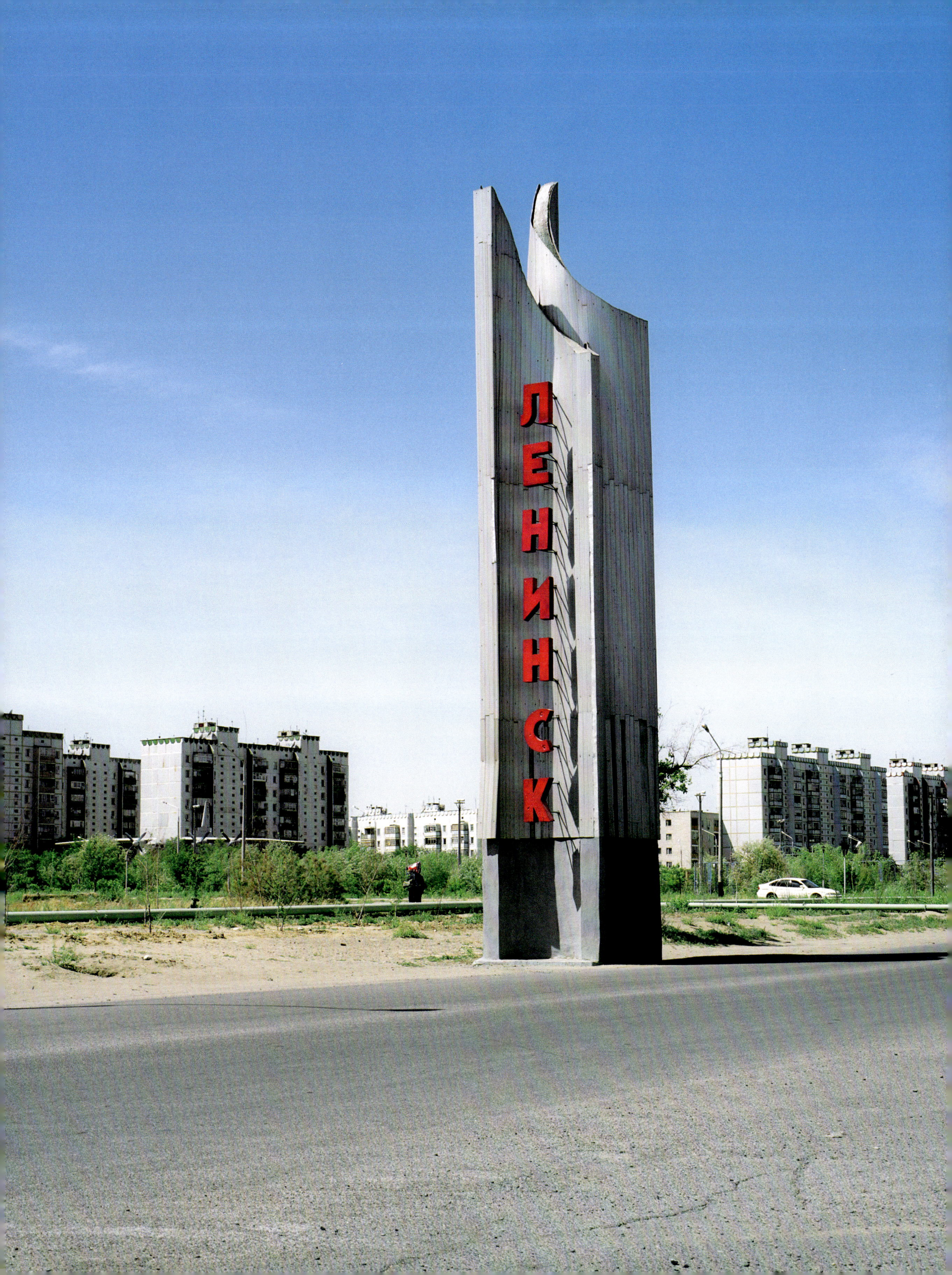
ЛЕНИНСК

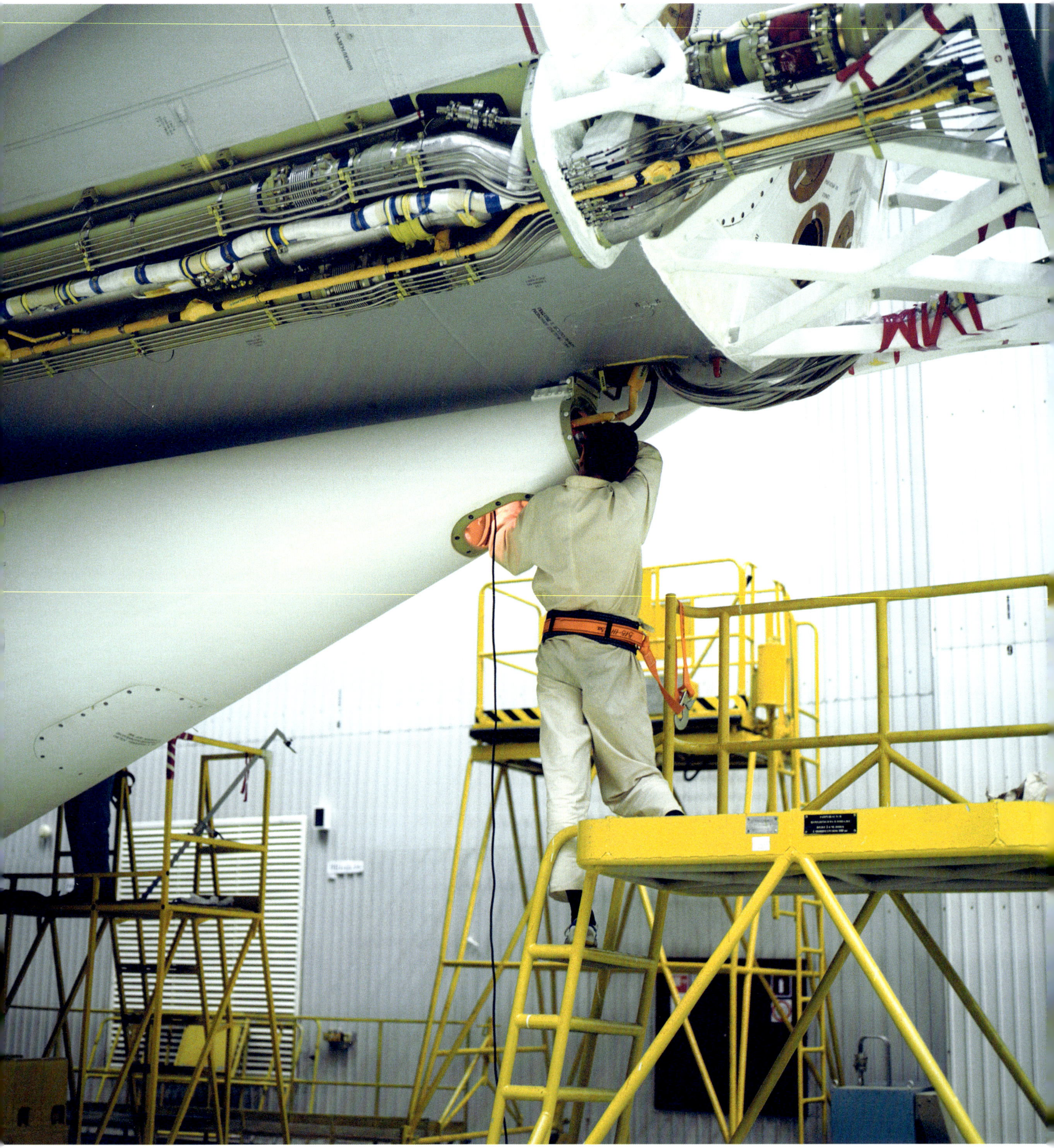

Engineer, Missile Assembly Building, Baikonur.

Far right: Students at the Missile Assembly Building.

Right: Postage stamps commemorating the Russian / Polish Soyuz–30 space flight (1978).

Below: Rocket support prop, Missile Assembly Building.

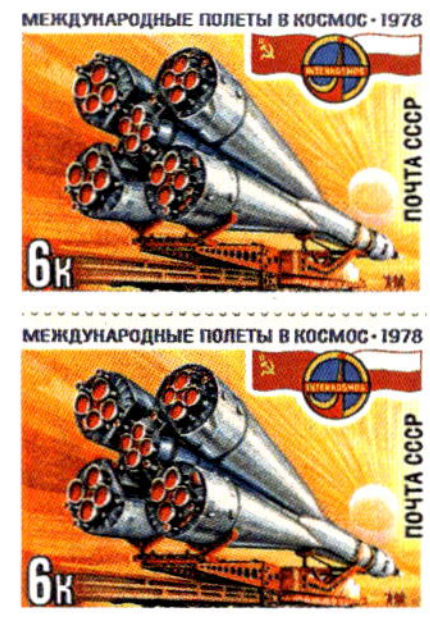

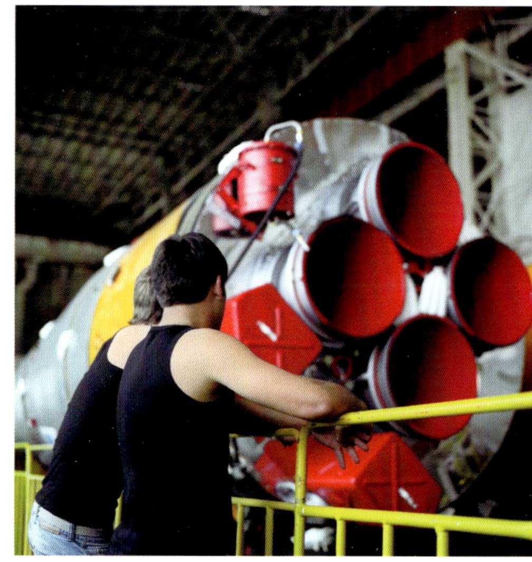

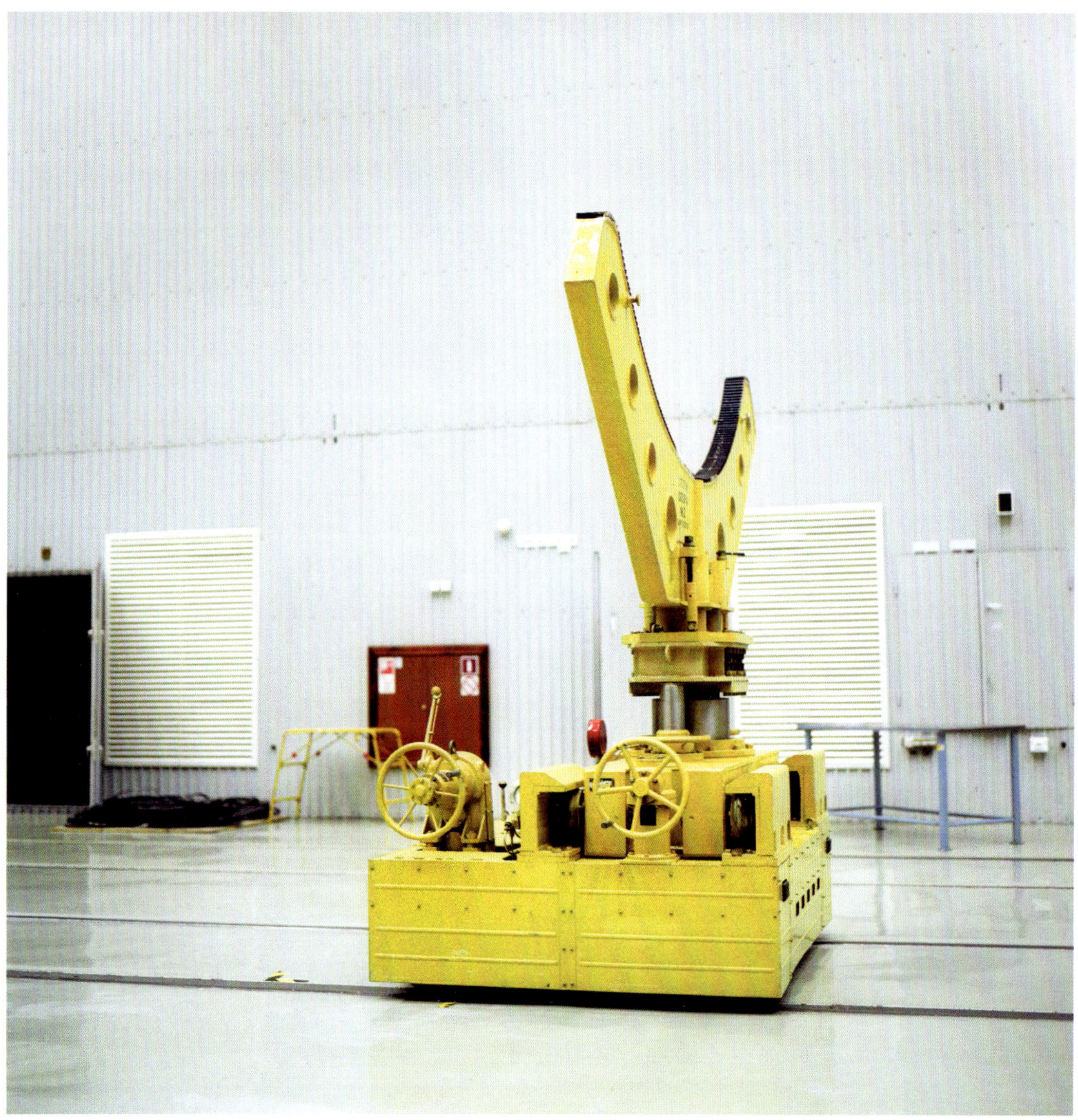

Engineer, Missile Assembly Building, Baikonur.

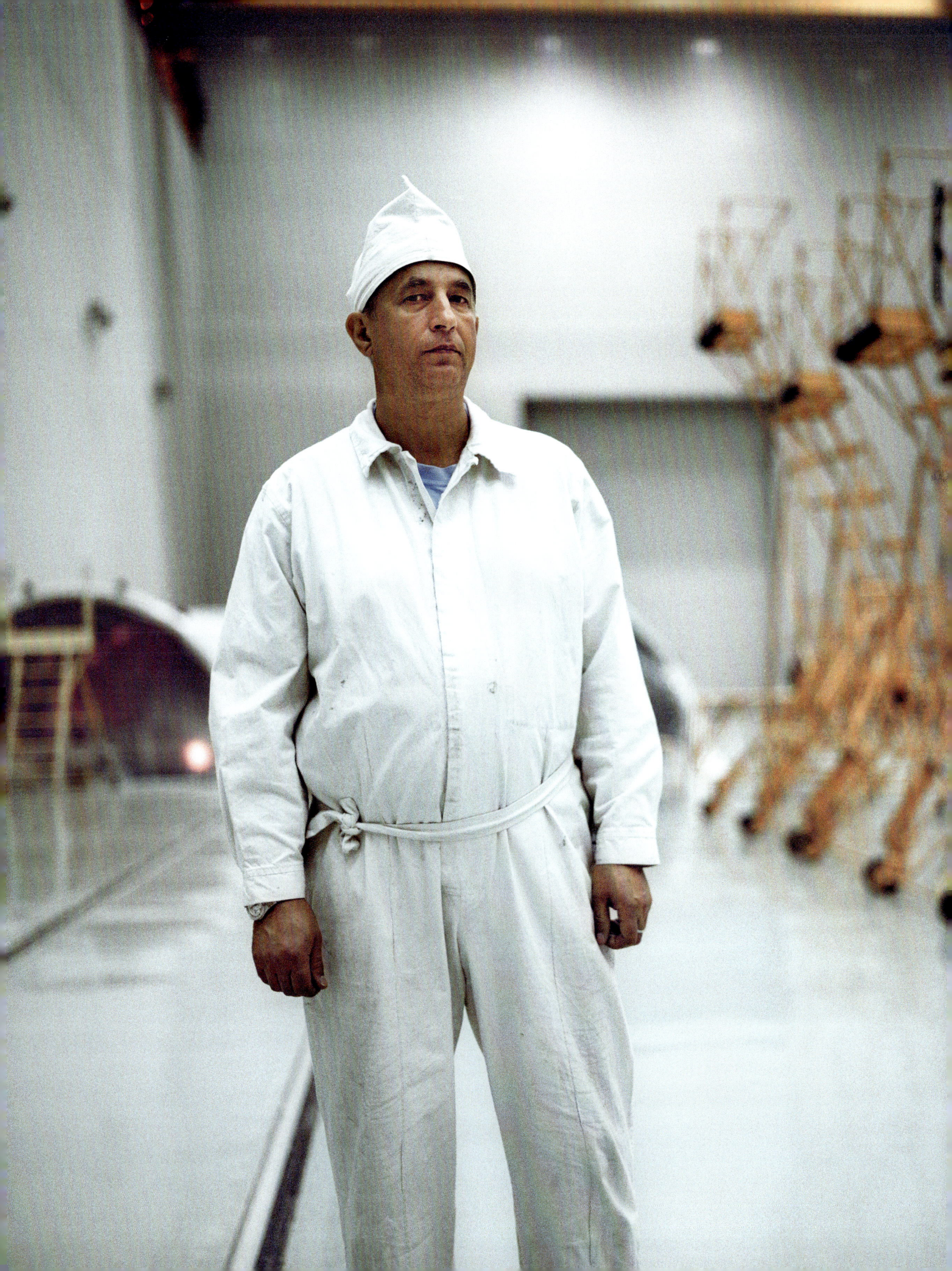

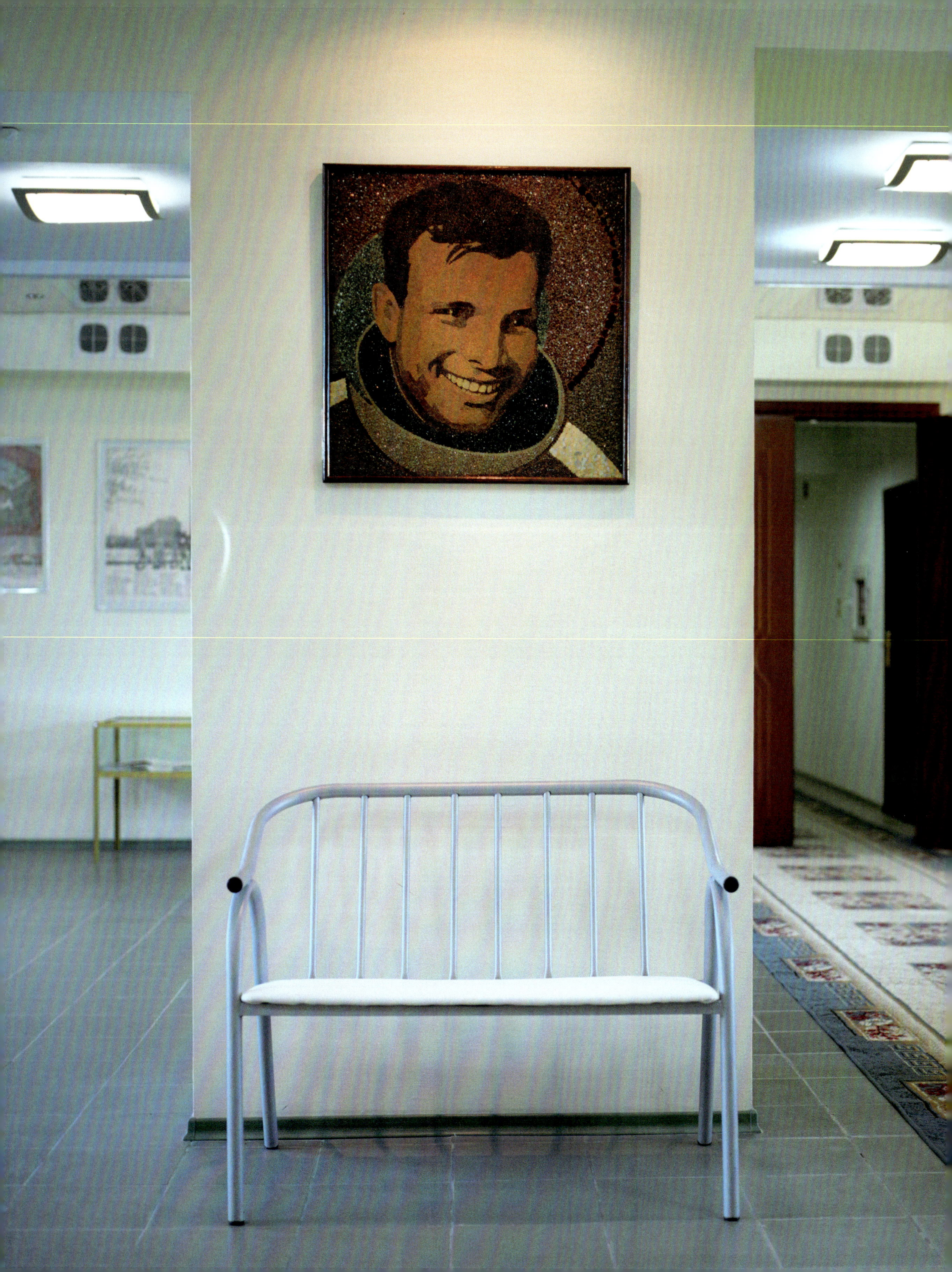

Portrait of Y A Gagarin made in semi-precious stones, Museum of Gagarin, Baikonur.

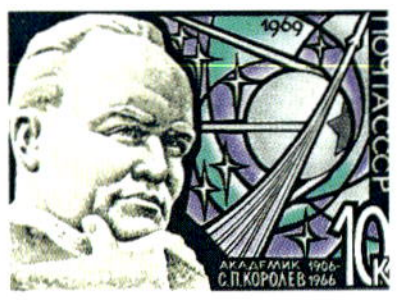

Right: Wall, Missile Assembly Building. The mural features S P Korolyov, and proclaims 'The road to the stars is open'.

Above: Postage stamp in memory of S P Korolyov, (1907–1966), the head Soviet rocket engineer, designer and constructor, and the founder of practical cosmonautics.

Below: Envelope commemorating Korolyov and signed by cosmonauts P R Popovich, A A Leonov, V F Bykovsky and V V Gorbatko.

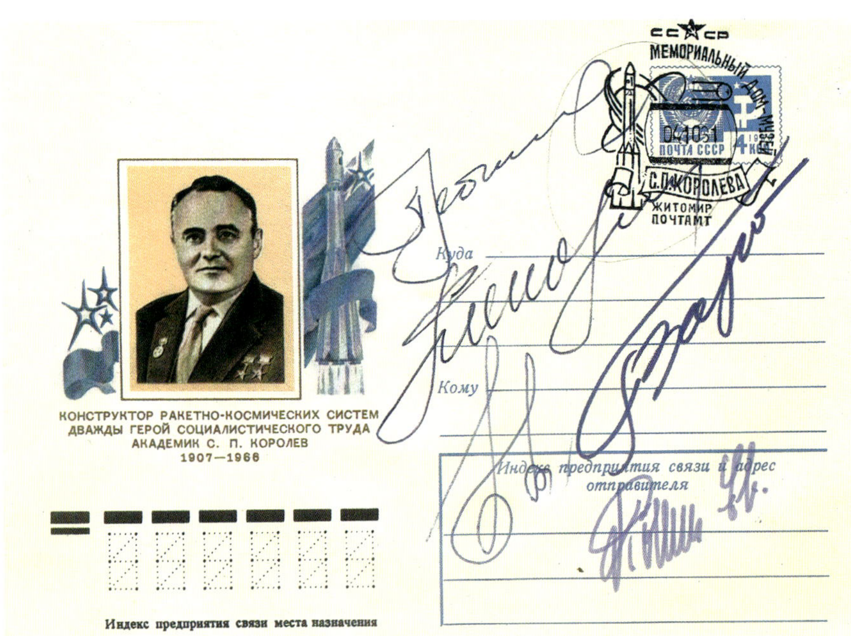

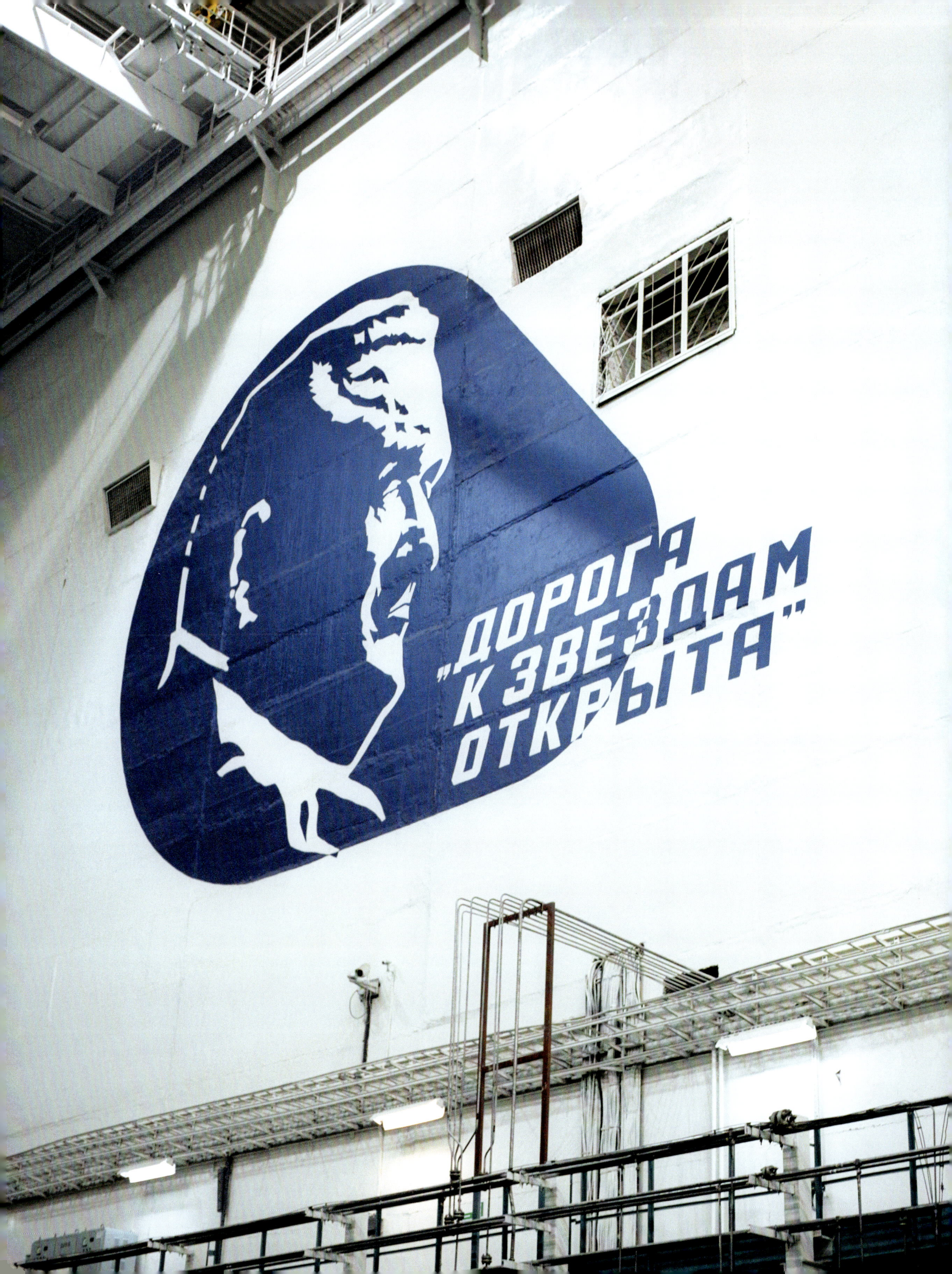
„ДОРОГА
К ЗВЕЗДАМ
ОТКРЫТА"

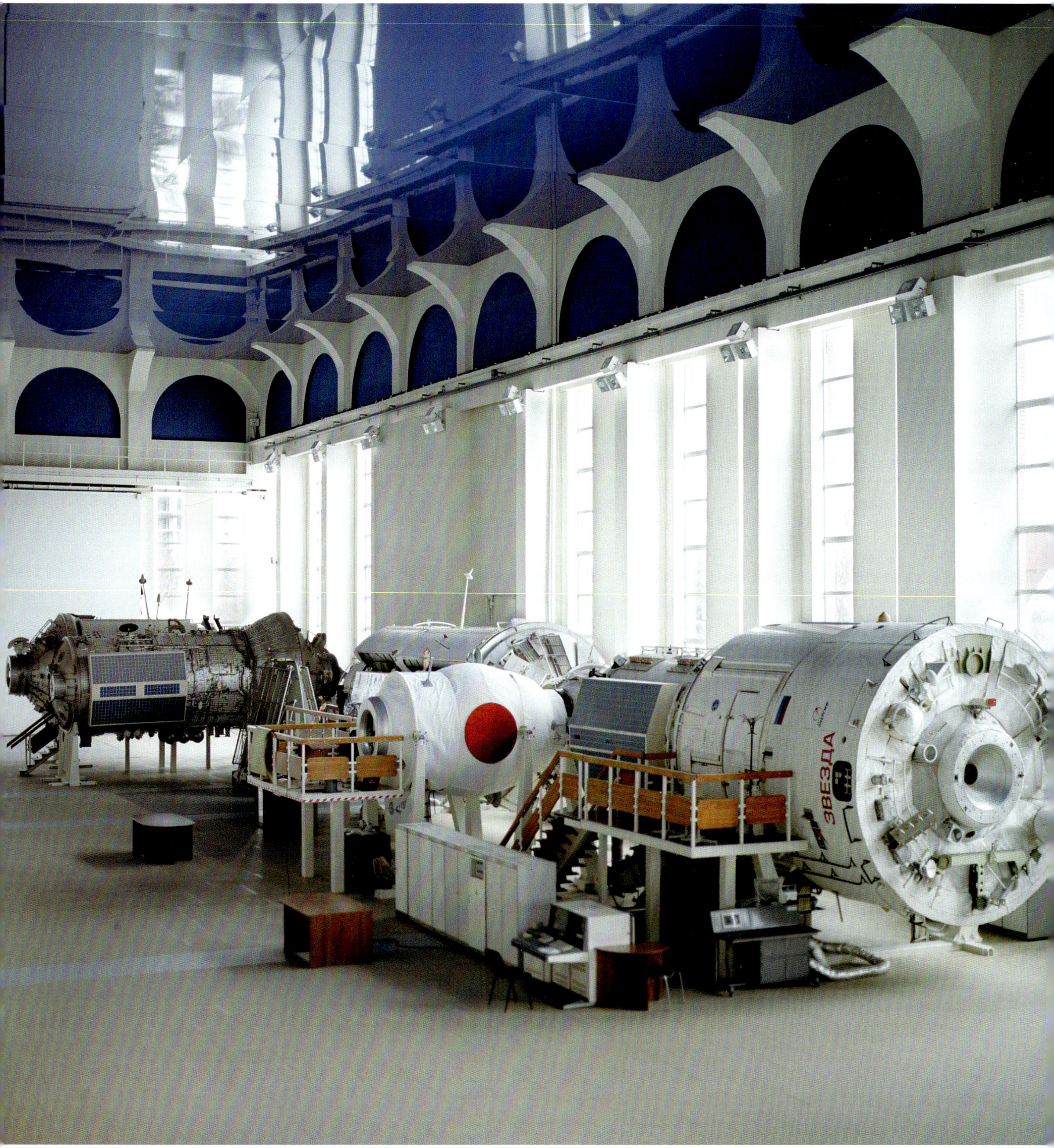

Zvezda service module, a component of the International Space Station (ISS).

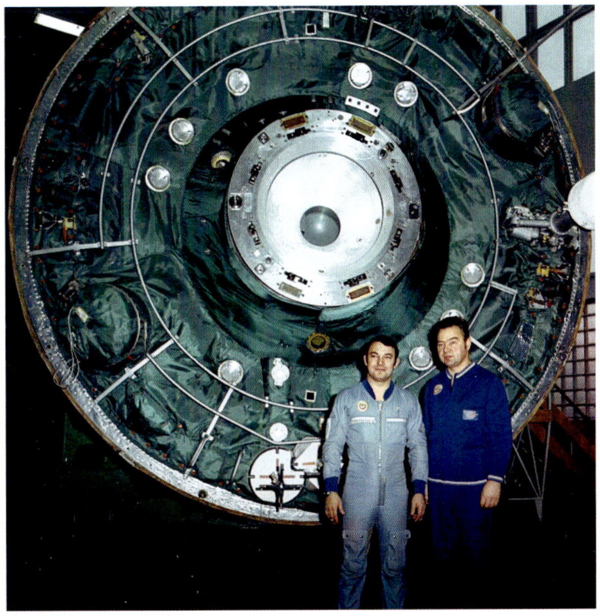

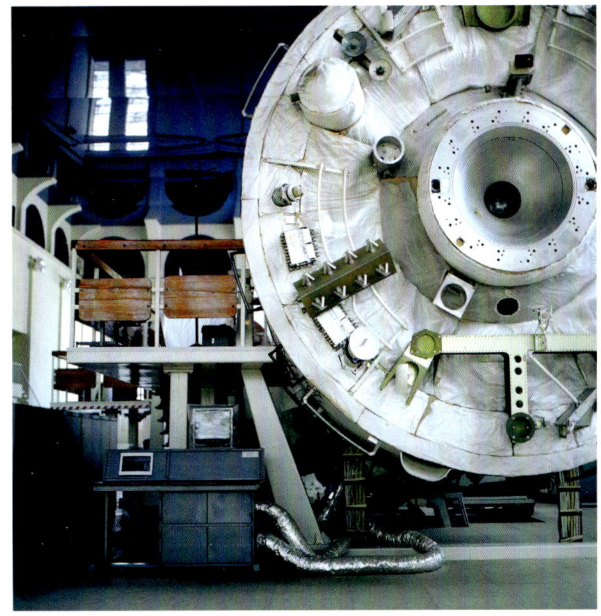

Above left: Cosmonauts, G M Grechko and Y V Romanenko, Soyuz–26 space flight crew, launched 16 January 1978.

Above right and left: Zvezda service module, a component of the ISS.

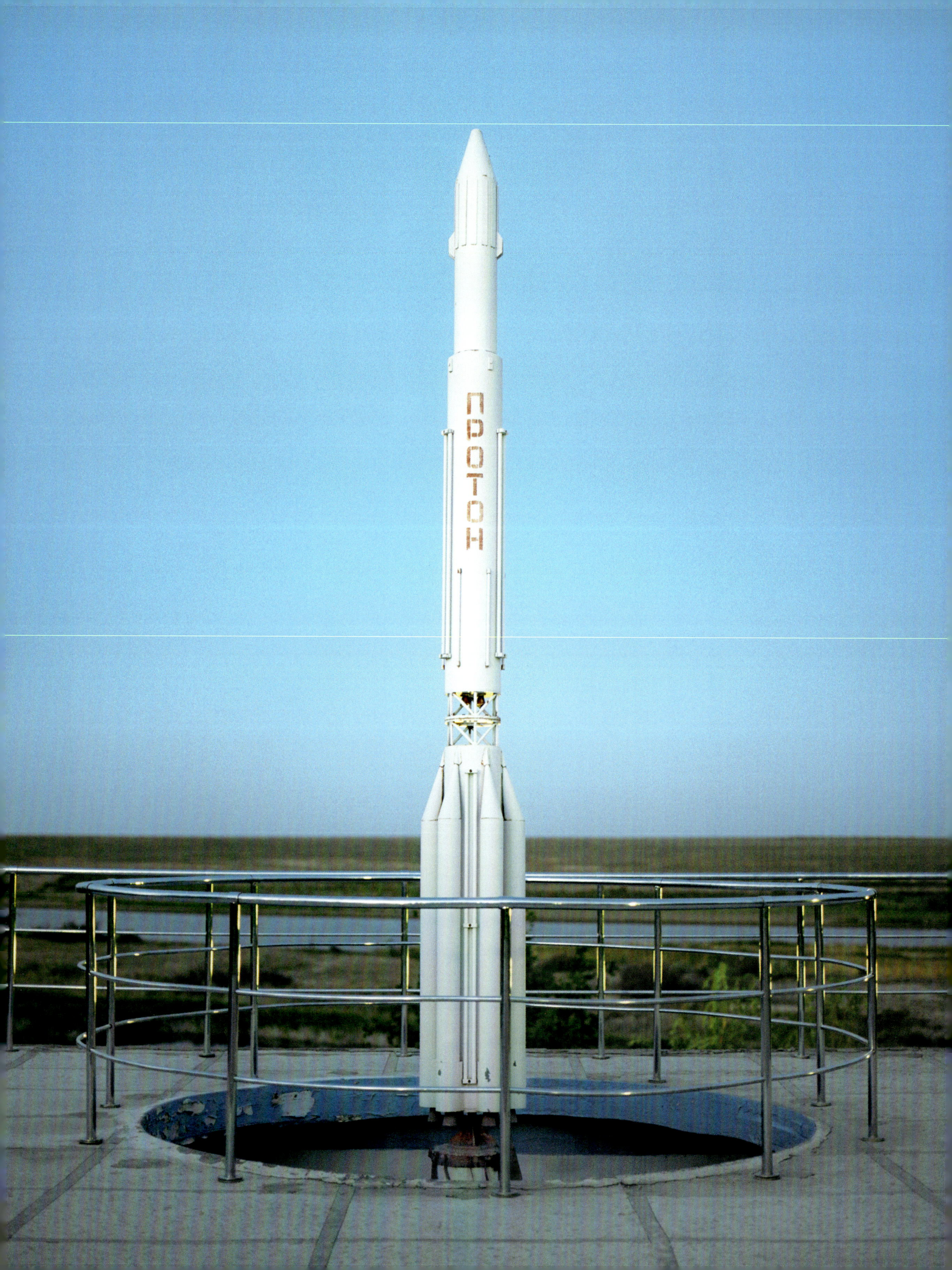
ПРОТОН

Left: Proton mock-up, an unmanned launch vehicle used for both commercial and Russian government space launches, Alley of Cosmonauts, Baikonur.

Below: Sculpture and toy shop, Baikonur.

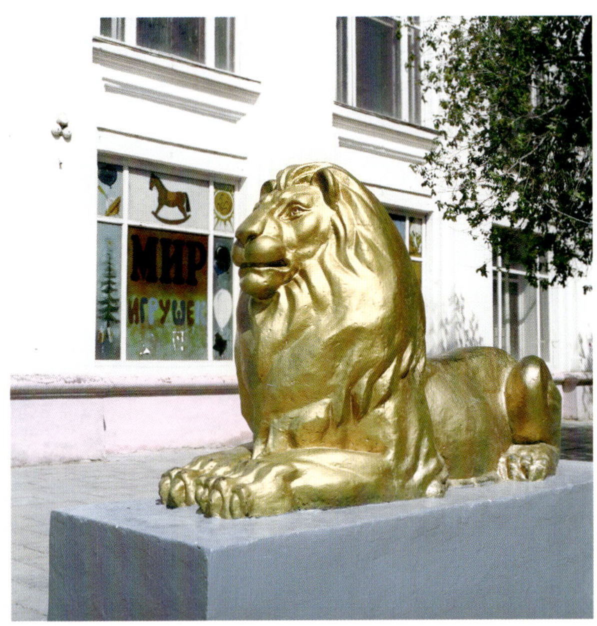

Obelisk, Baikonur.

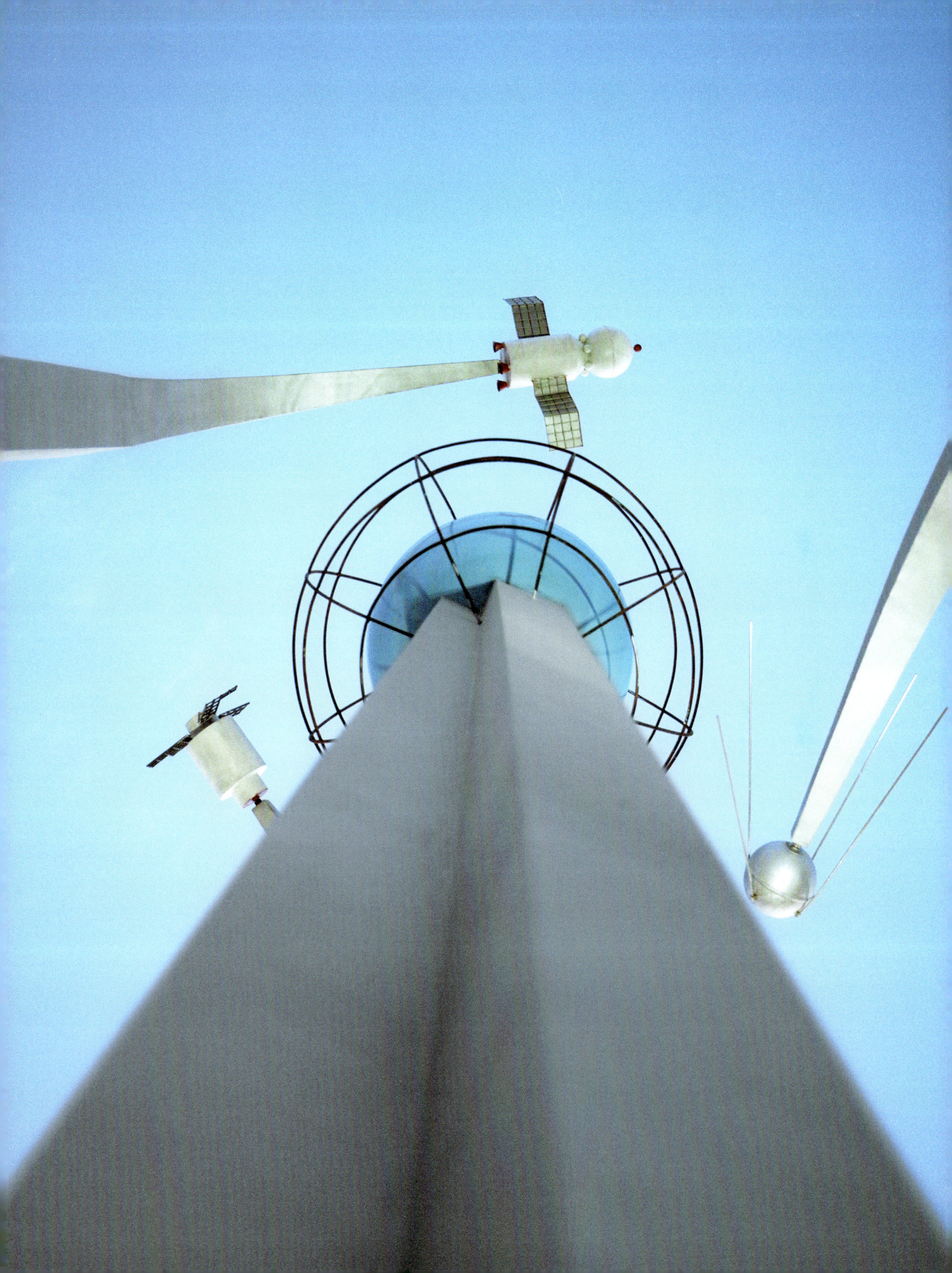

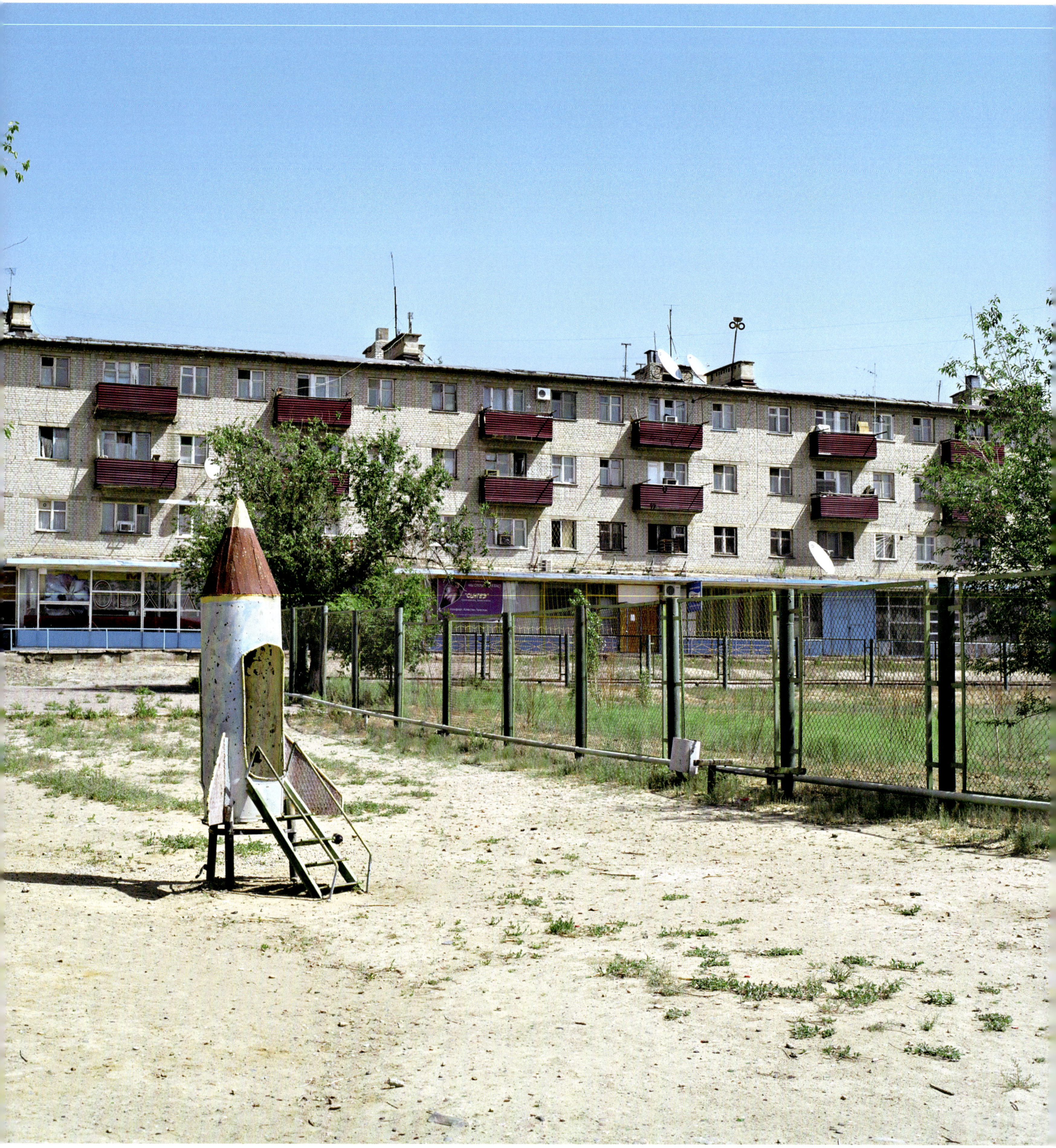

Playground, Baikonur.

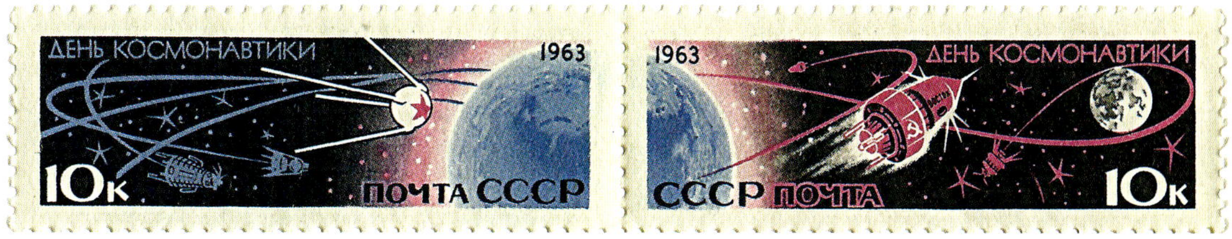

Double postage stamp (1963), commemorating Cosmonautics Day.

Obelisk commemorating the launch of Sputnik-1,
the first Earth-orbiting artificial satellite, Baikonur.

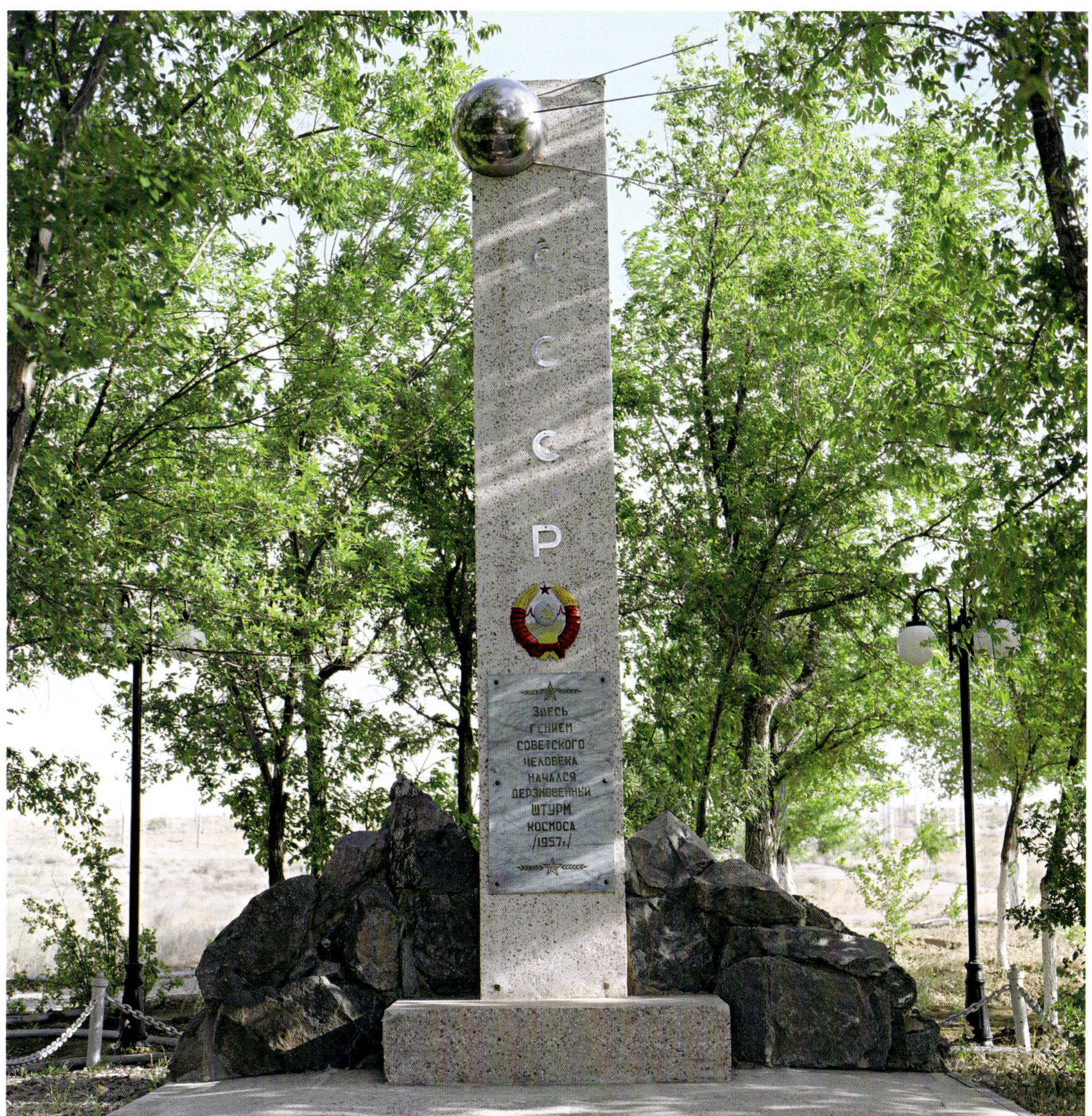

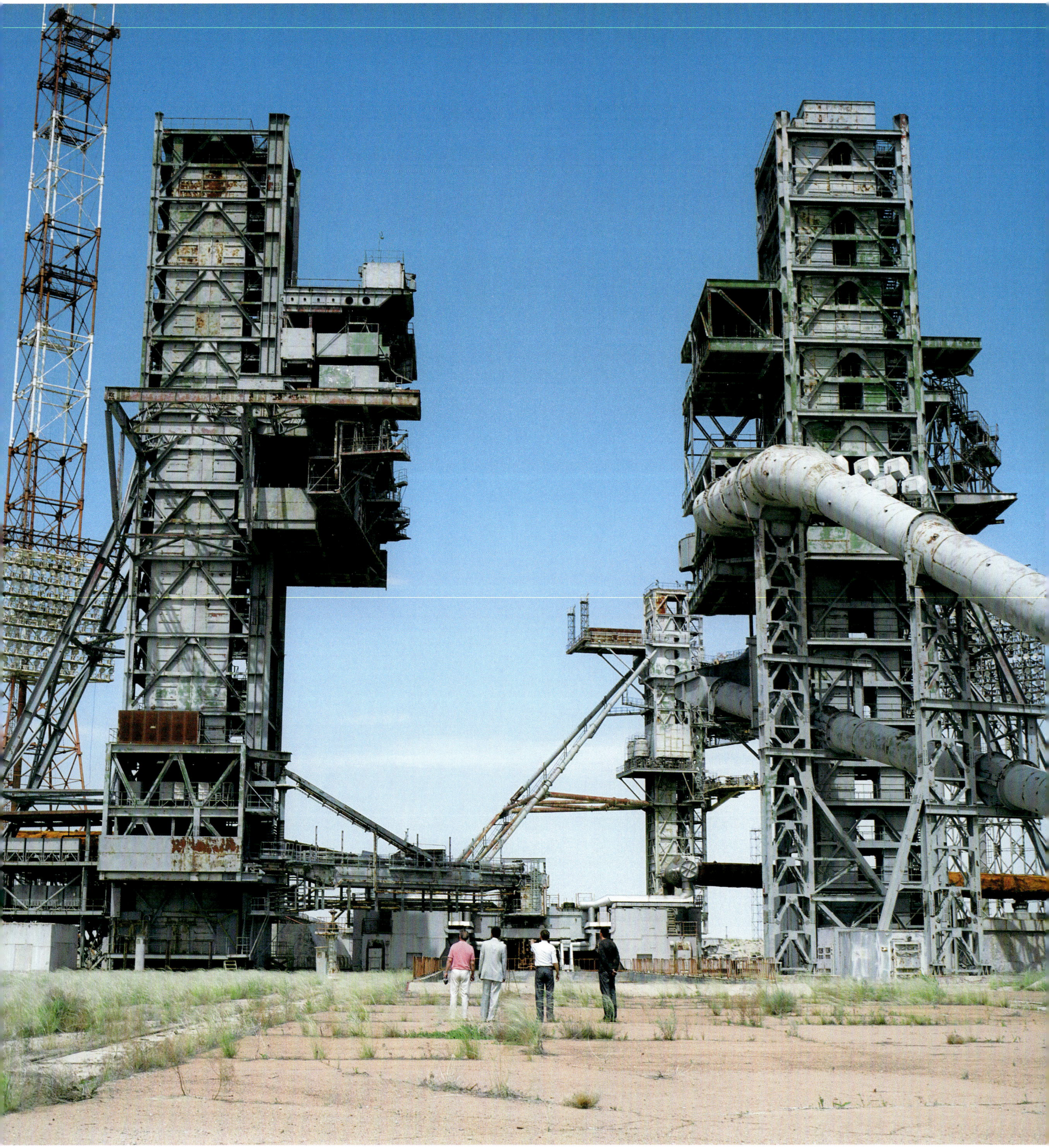

Abandoned Buran launch site.

Entrance to the abandoned Buran launch site. Buran spacecraft, analog of the American Shuttle programme, completed only one space flight before the cancellation of the Soviet Shuttle programme in 1993. Cosmonauts, however, continued preparation for the Buran programme until 1995. They were then transfered to the Mir preparation programme.

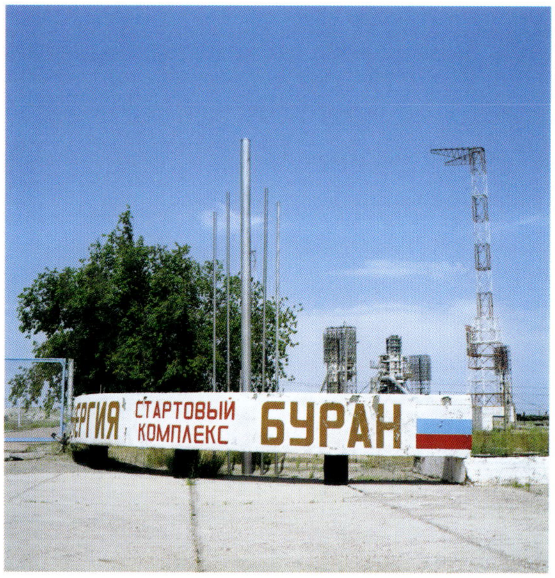

Cockpit of the Buran spacecraft, installation, Museum of Gagarin, Baikonur.

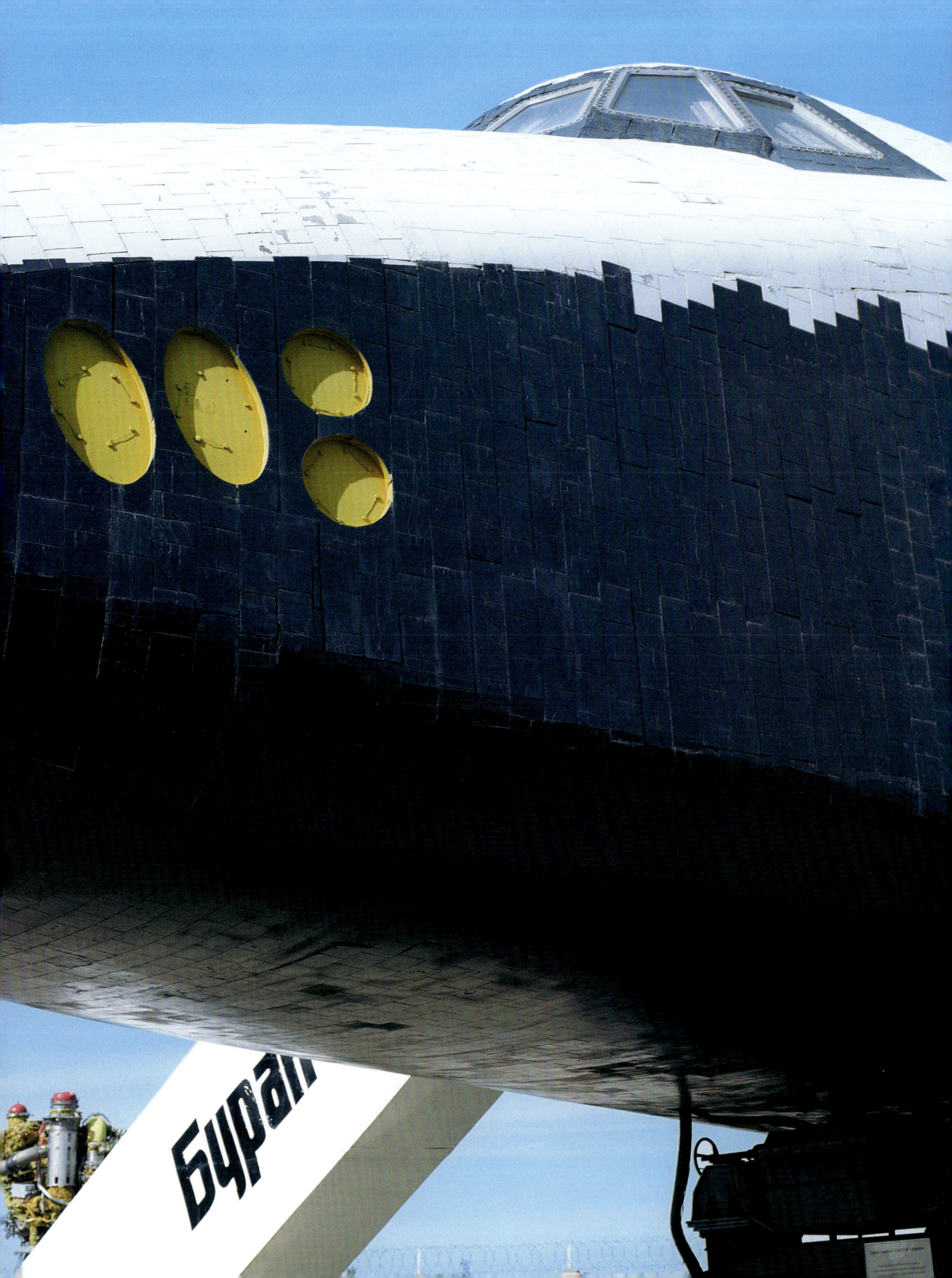

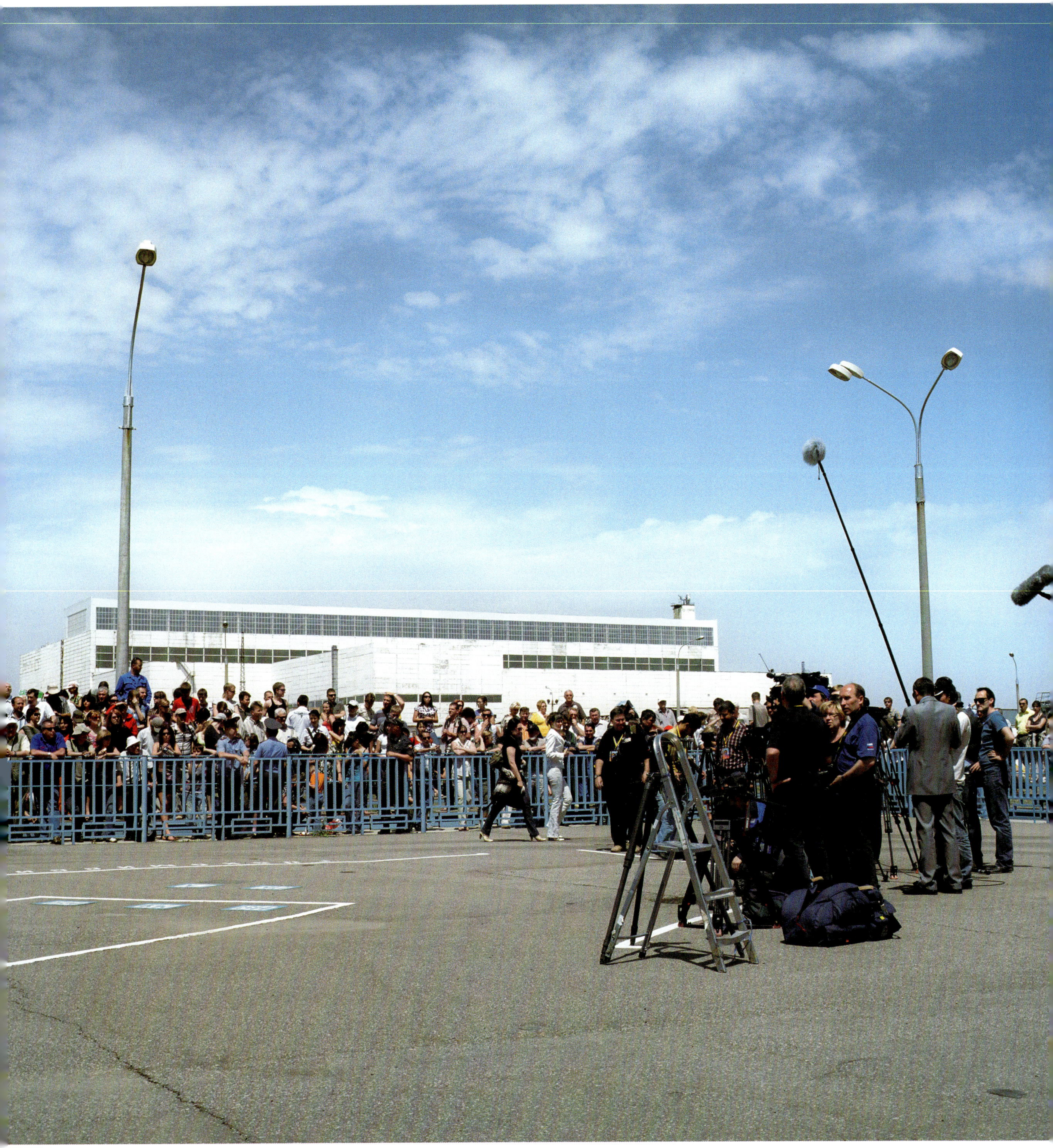

The press waiting for cosmonauts.

Right: Cosmonauts, R Y Romanenko, Frank De Winne (ESA astronaut), Robert Thirsk (CSA astronaut) on their way to make their final official report to Dr. Prof A N Perminov.

Bottom left: Cosmonaut, P I Klimuk, crew member of Soyuz–13 space flight, launched 18 December 1973.

Bottom right: Bus ready to take cosmonauts to launch pad.

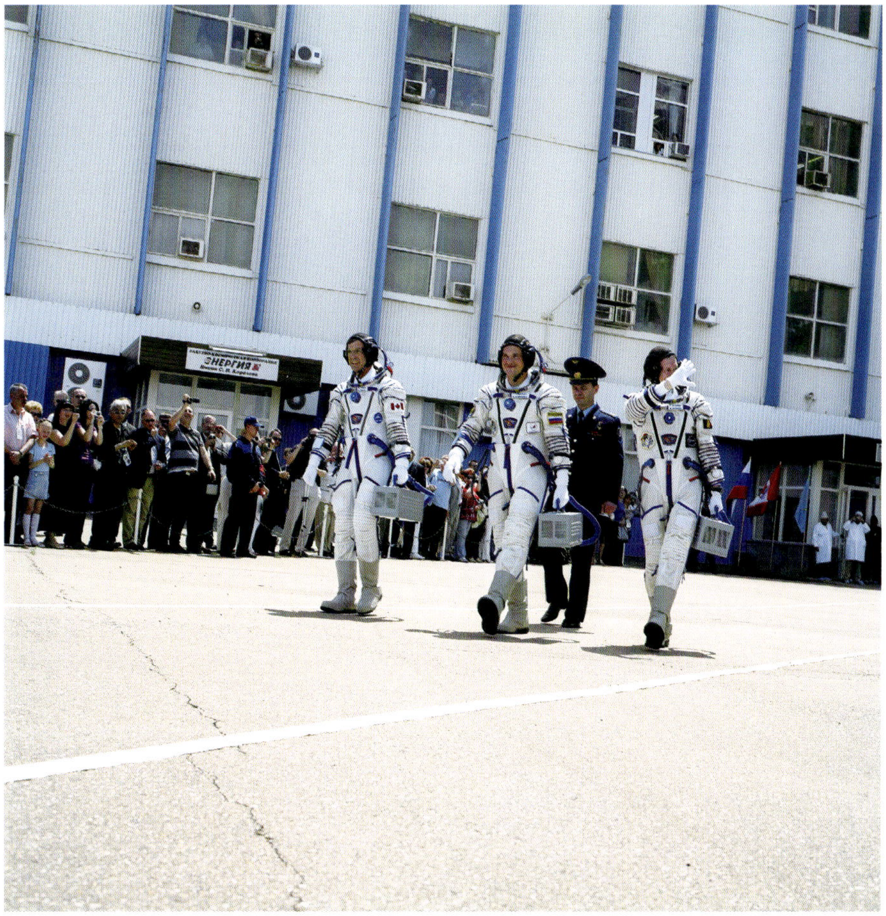

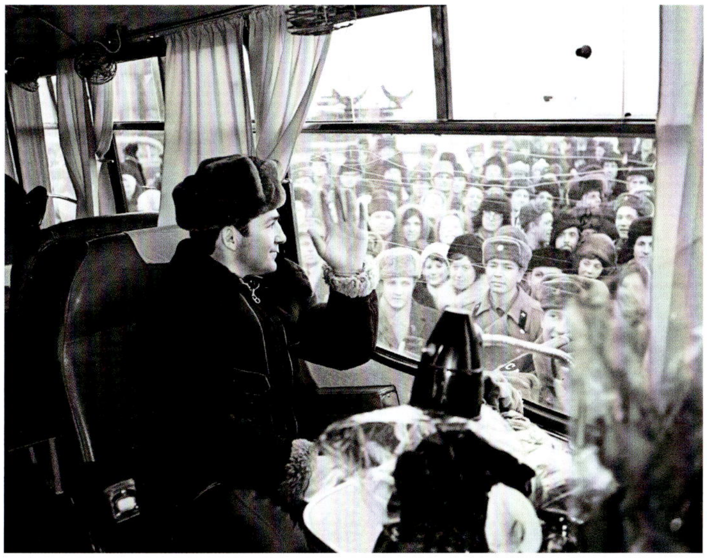

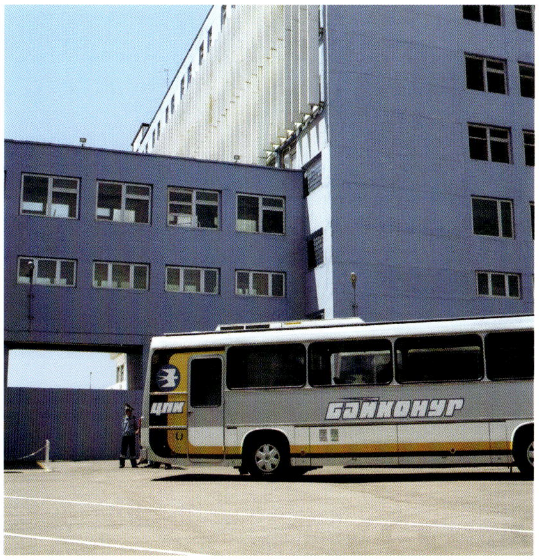

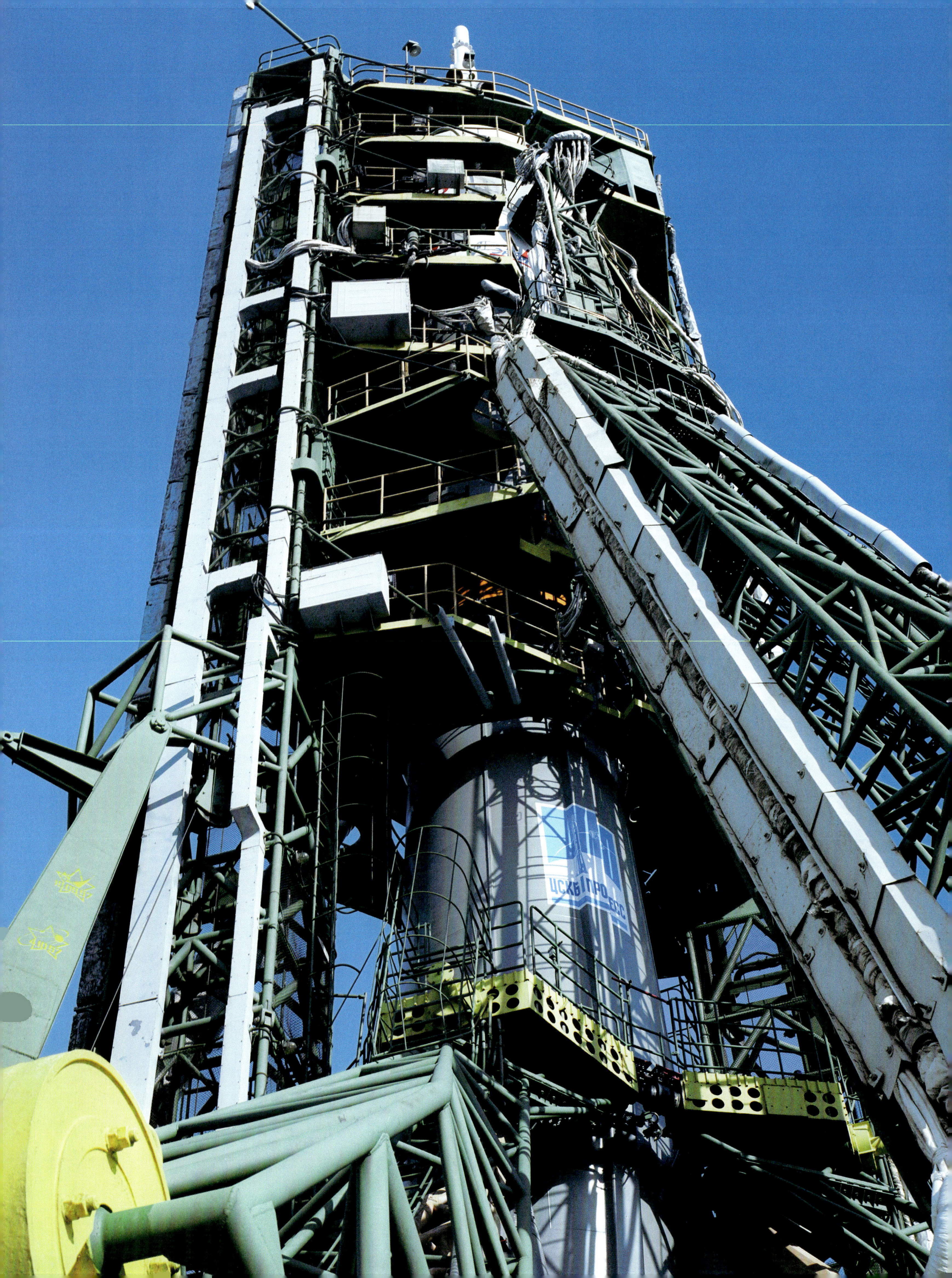

Soyuz spacecraft on the launch pad.

Launching of Soyuz TMA–15 space mission

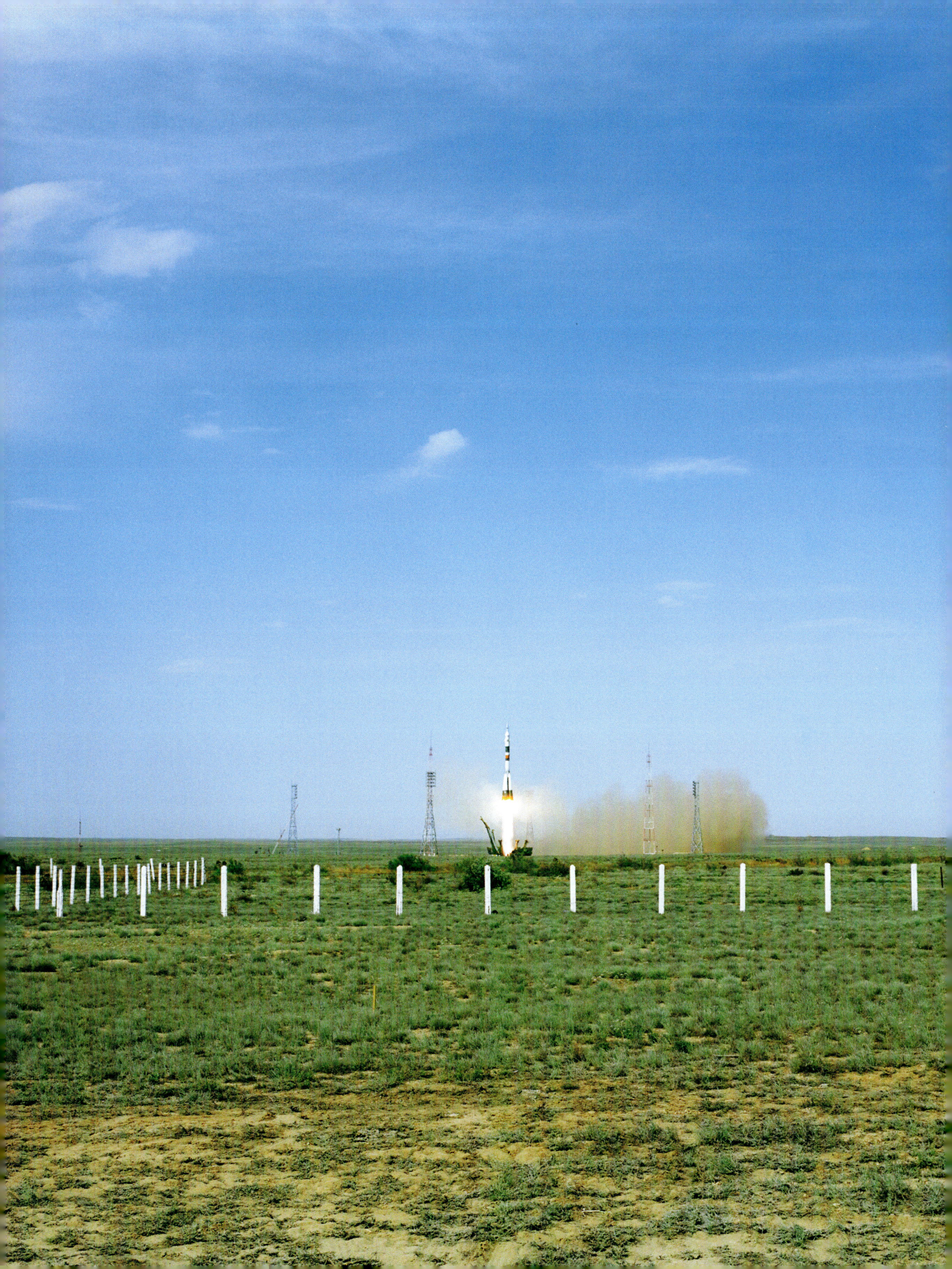

INDEX OF PHOTOGRAPHS

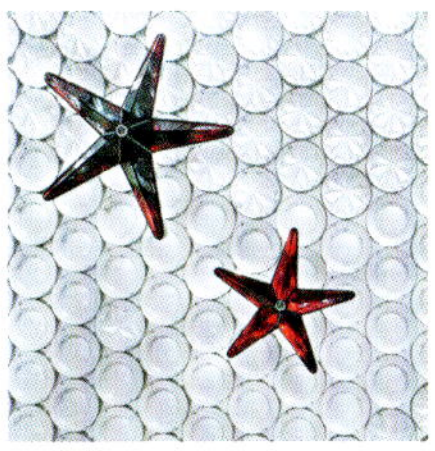

Decorative mosaic, detail,
Main headquarters building,
Star City.
Photograph, 2010.

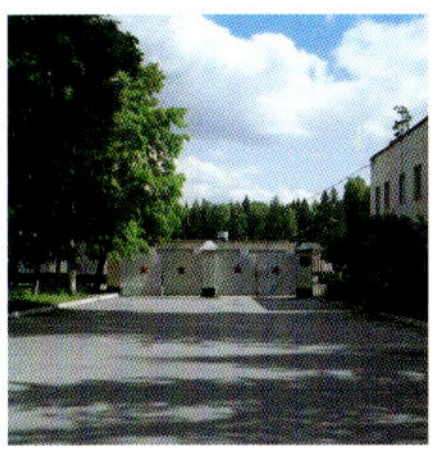

Gates, garage, petrol station
and repair centre, Star City.
Photograph, 2010.

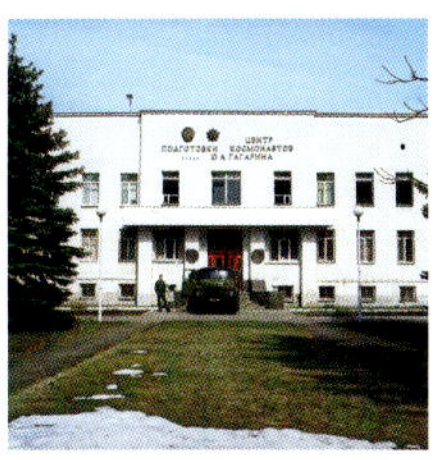

Main headquarters
building, Star City.
Photograph, 2010.

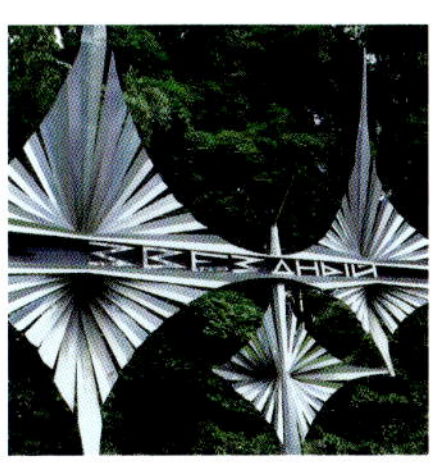

Star City road sign.
Photograph, 2010.

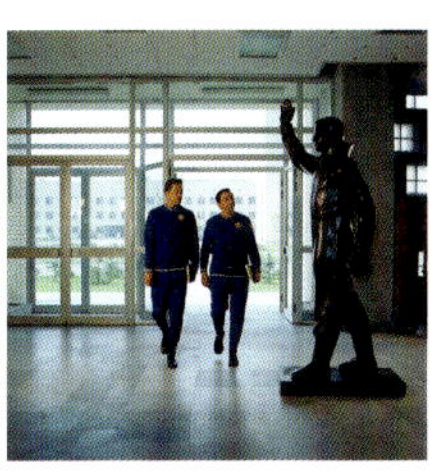

Cosmonauts,
Bykovsky and Aksyonov,
2nd Department, Star City.
Archival photograph, 1976.

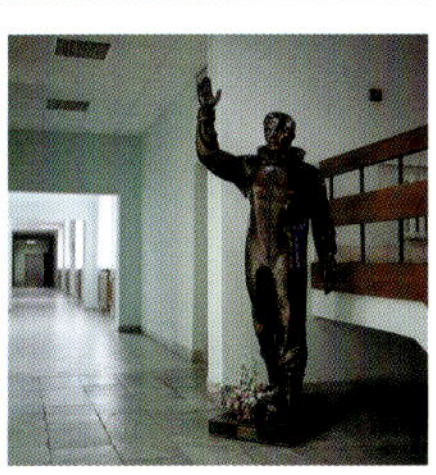

Monument to Gagarin,
2nd Department, Star City.
Photograph, 2010.

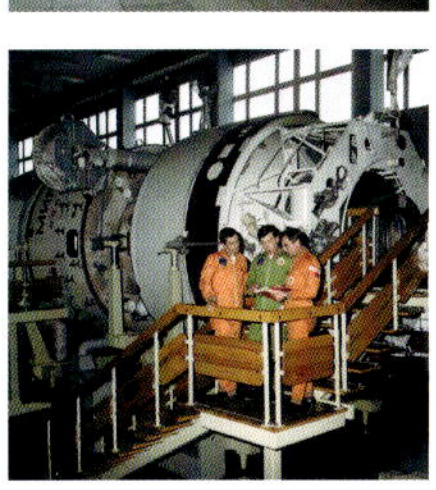

Cosmonauts, Volkov,
Krikalev, J-L M Chrétien,
Mir space station,
2nd Department, Star City.
Archival photograph, 1989.

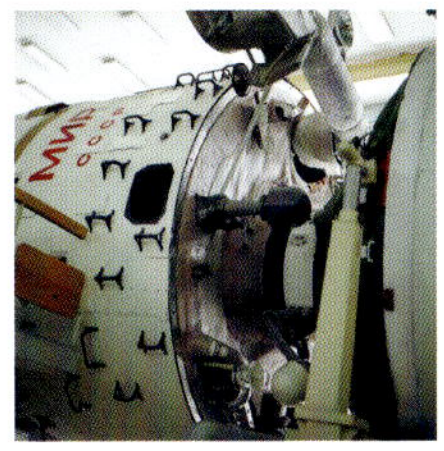

Mir space station,
2nd Department, Star City.
Photograph, 2009.

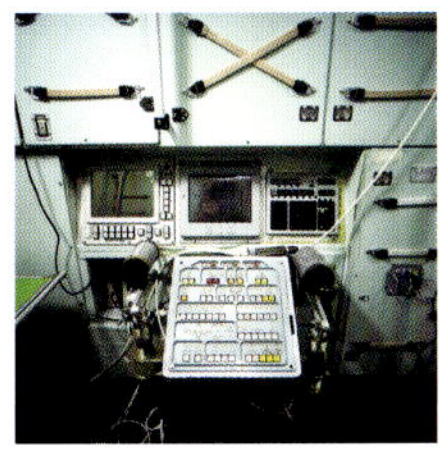

Teleoperator's manual
control panel, Mir space
station, 2nd Department,
Star City.
Photograph, 2010.

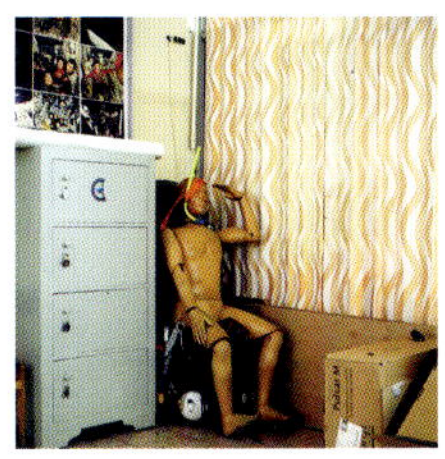

Unused crash dummy,
3rd Department, Star City.
Photograph, 2010.

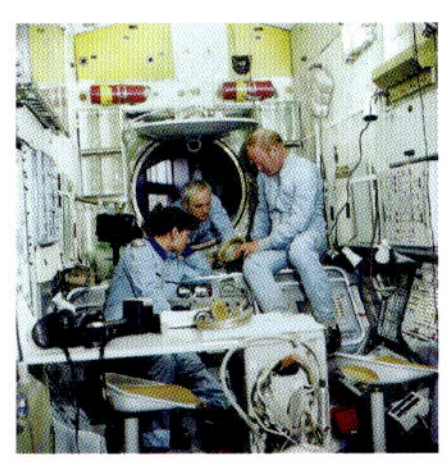

Cosmonauts, Savitskaya,
Dzhanibekov, Volk,
Mir space station,
2nd Department, Star City.
Archival photograph, 1984.

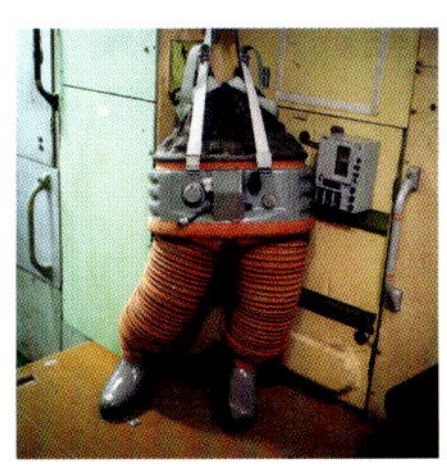

Negative pressure suit
(ODNT), Mir space station,
2nd Department, Star City.
Photograph, 2010.

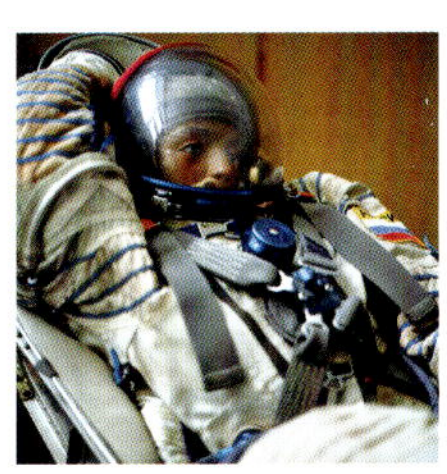

Dummy wearing an
outdated space suit,
2nd Department, Star City.
Photograph, 2009.

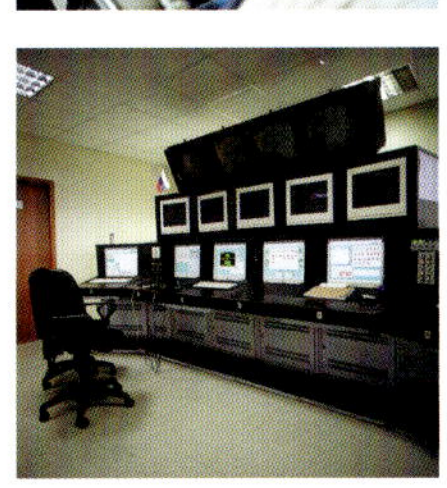

Control room,
TDK–7ST simulator,
2nd Department, Star City.
Photograph, 2010.

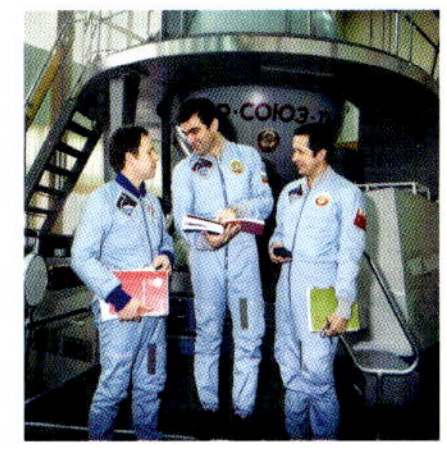

Cosmonauts, Solovyov,
Savinykh, Aleksandrov,
2nd Department, Star City.
Archival photograph, 1988.

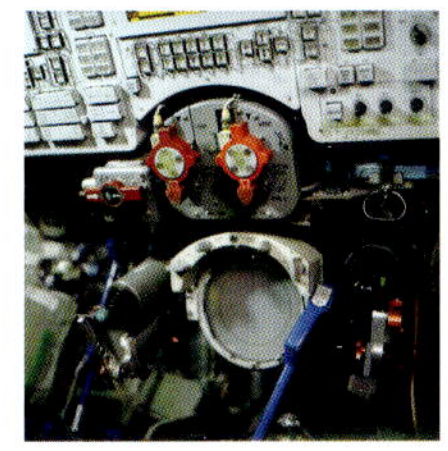

Control panel,
Soyuz–TM simulator,
2nd Department, Star City.
Photograph, 2010.

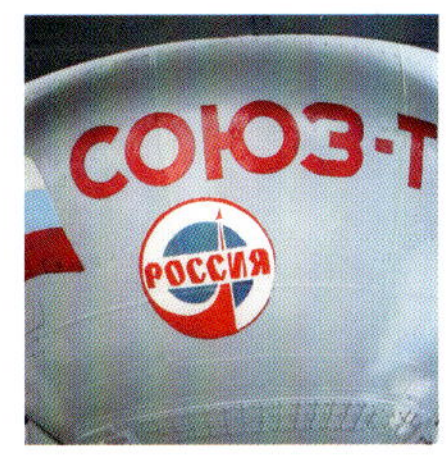

Soyuz–TM simulator,
2nd Department, Star City.
Photograph, 2010.

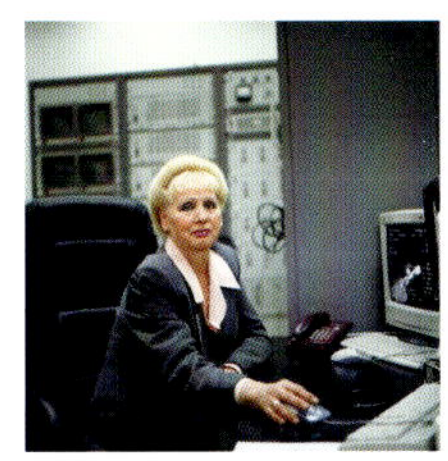

Engineer, Osipova,
TDK–7ST simulator
control panel,
2nd Department, Star City.
Photograph, 2010.

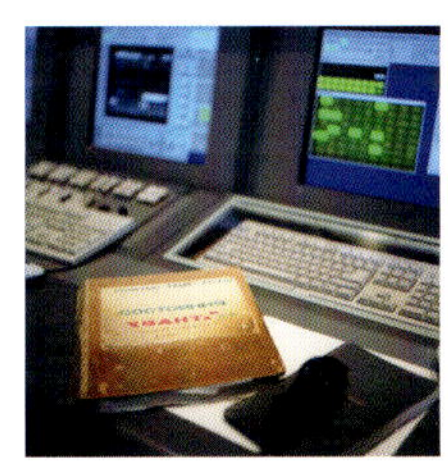

Control panel,
TDK–7ST simulator,
2nd Department, Star City.
Photograph, 2010.

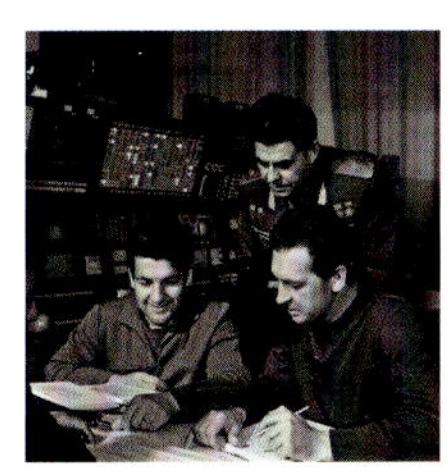

Cosmonauts,
Klimuk, Lebedev.
Archival photograph, 1973.

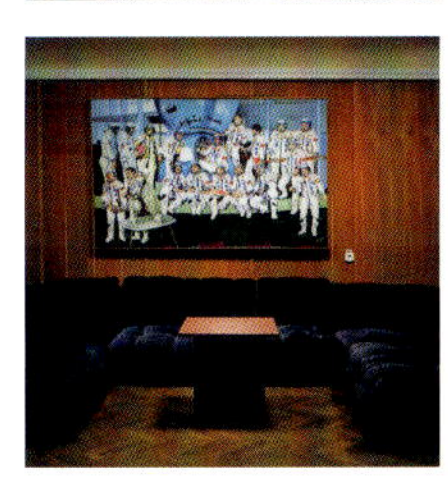

Interkosmos painting,
R Bobovich, 1982.
Reception area,
2nd Department, Star City.
Photograph, 2010.

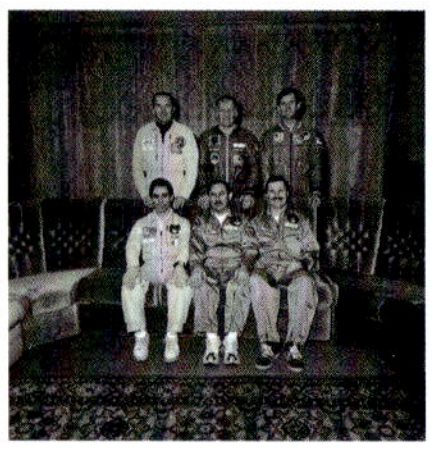

Main and back-up crew,
Soyuz TM–27 space flight,
Baikonur.
Archival photograph, 1998.

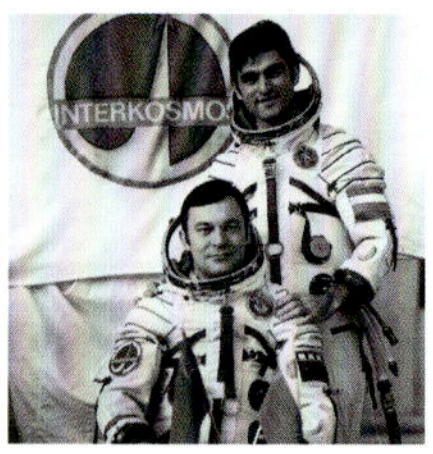

Cosmonauts,
Romanenko, A T Mendez.
Archival photograph, 1980.

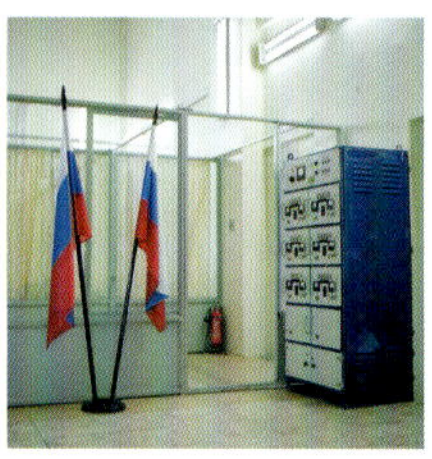

Redundant fuse panel,
3rd Department, Star City.
Photograph, 2010.

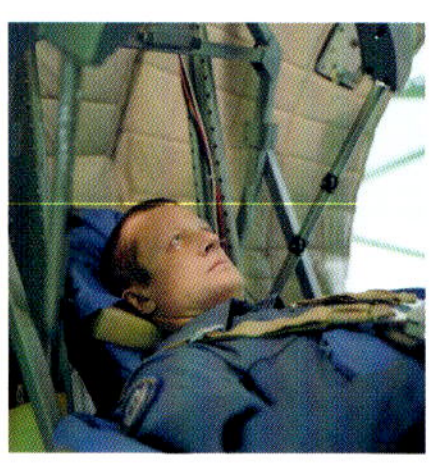

Cosmonaut, Kondratyev,
TsF–7 dynamic
trainer centrifuge,
3rd Department, Star City.
Photograph, 2010.

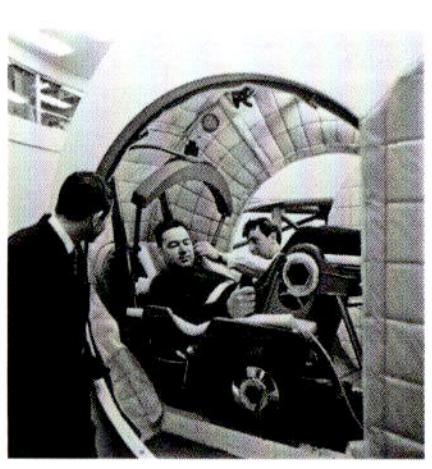

Cosmonaut, Grechko,
TsF–7 dynamic trainer
centrifuge,
3rd Department, Star City.
Archival photograph, 1980.

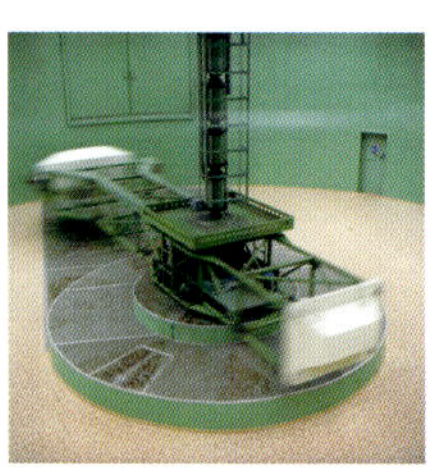

TsF–7 dynamic trainer
centrifuge,
3rd Department, Star City.
Photograph, 2010.

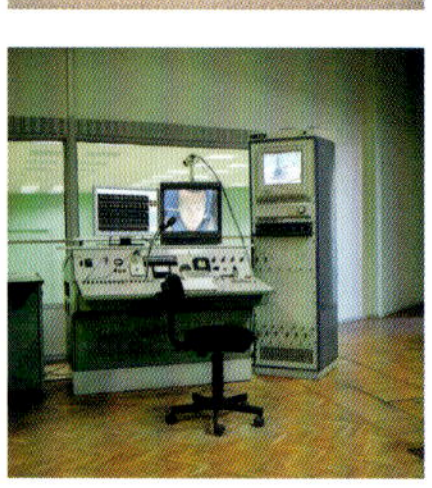

Medical control panel,
TsF–7 dynamic trainer
centrifuge,
3rd Department, Star City.
Photograph, 2010.

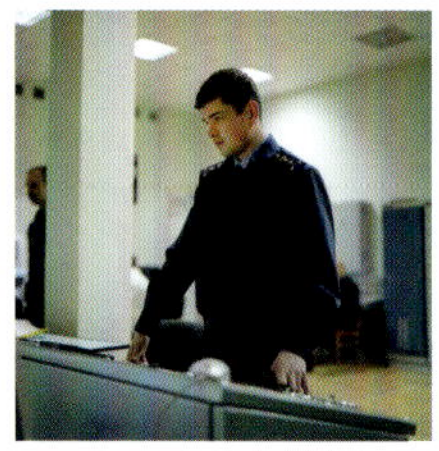

Engineer, Sergeev,
TsF–7 dynamic trainer
centrifuge, control panel,
3rd Department, Star City.
Photograph, 2010.

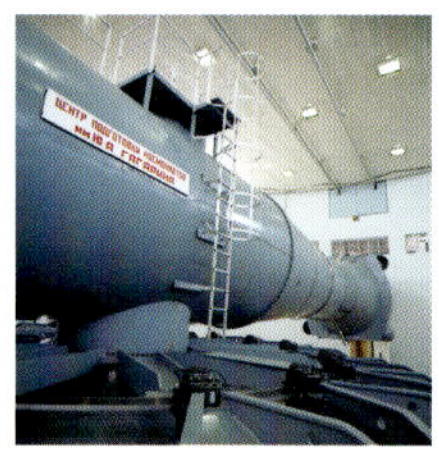

TsF–18 centrifuge
simulator,
3rd Department, Star City.
Photograph, 2010.

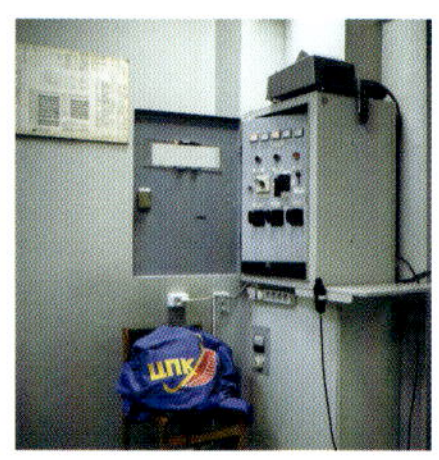

TsF–18 centrifuge
simulator's step platform
remote control panel,
3rd Department, Star City.
Photograph, 2010.

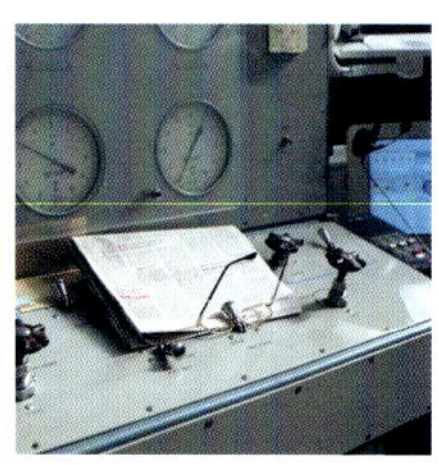

TsF–18 centrifuge
simulator, control panel,
3rd Department, Star City.
Photograph, 2010.

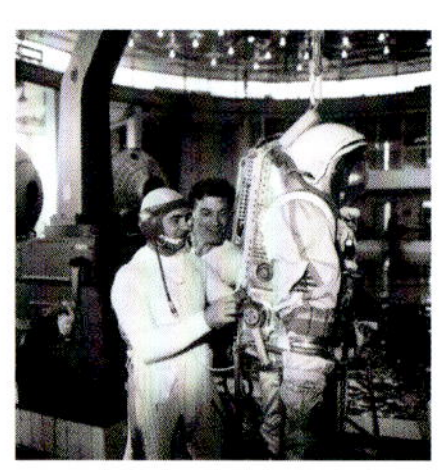

Cosmonauts,
Viktorenko and Serebrov,
Hydro laboratory, Star City.
Archival photograph.

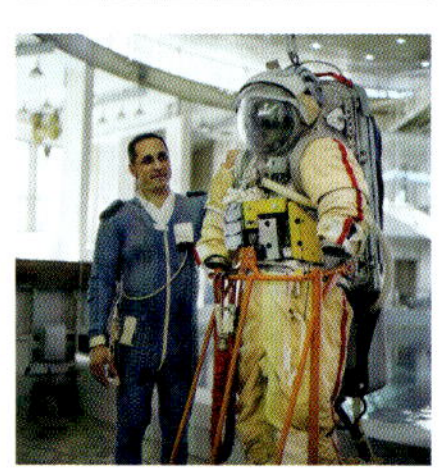

Cosmonaut, Shkaplerov
and hydro-suit,
Hydro laboratory, Star City.
Photograph, 2010.

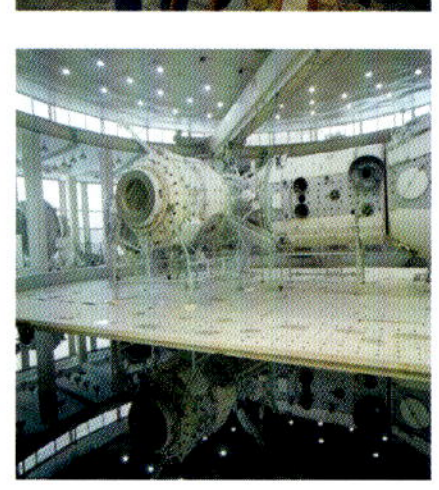

ISS mock-up,
Hydro laboratory, Star City.
Photograph, 2010.

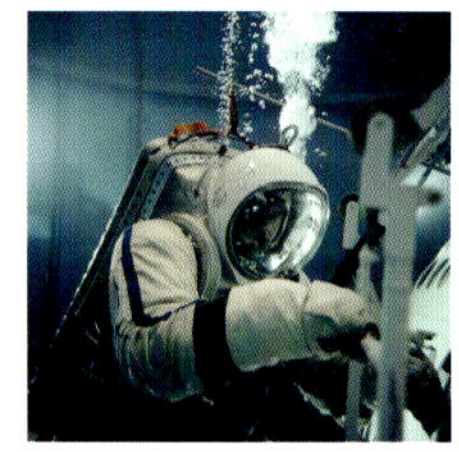

Training session,
Hydro laboratory, Star City.
Archival photograph, 1984.

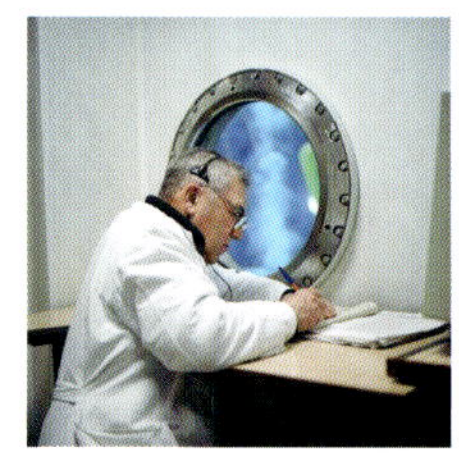

Medic, Krasnov,
Hydro laboratory, Star City.
Photograph, 2010.

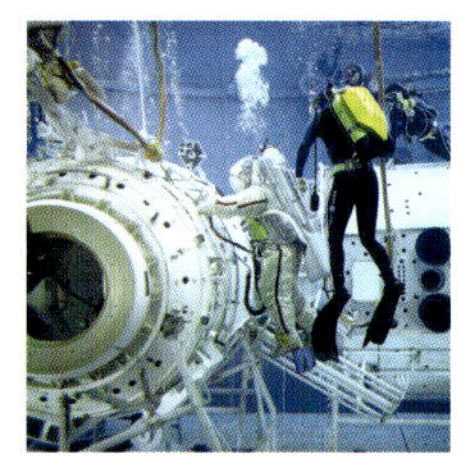

Cosmonaut, Shkaplerov,
Hydro laboratory, Star City.
Photograph, 2010.

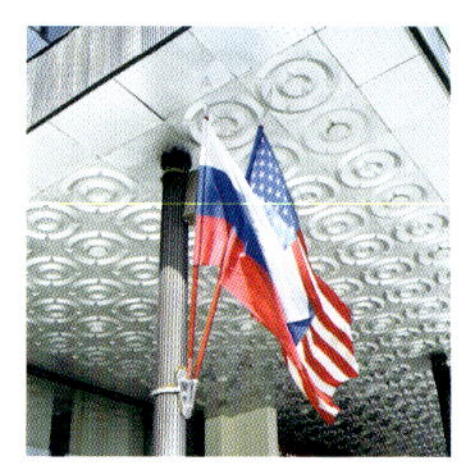

Flags, House of
Cosmonauts, Star City.
Photograph, 2009.

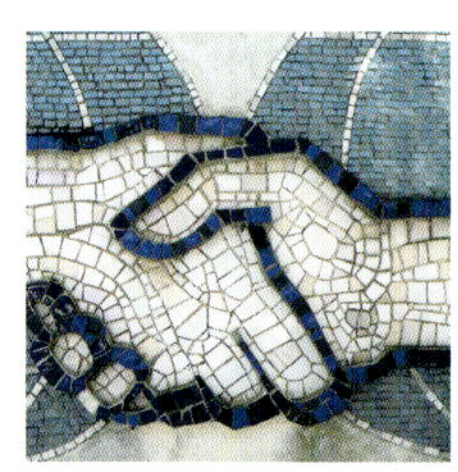

Decorative mosaic, detail,
Star City.
Photograph, 2010.

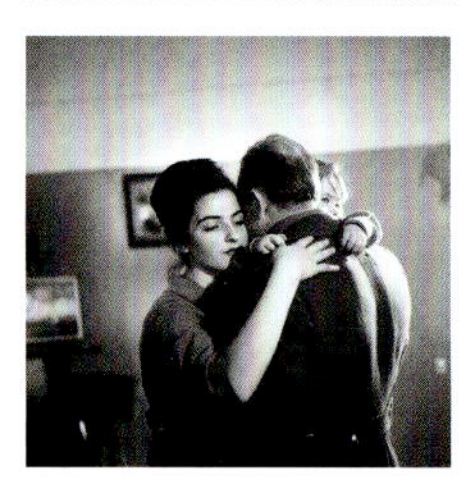

Cosmonaut, Leonov,
his wife and daughter,
Star City.
Archival photograph.

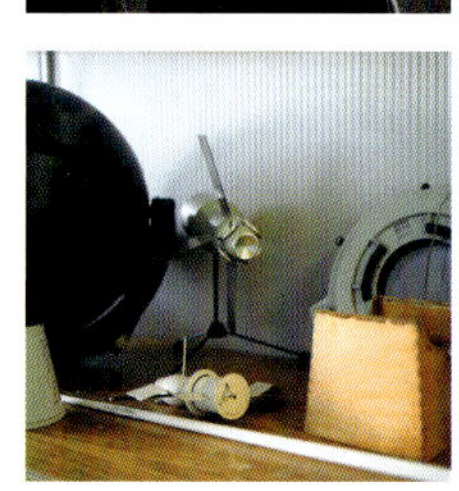

Mock-ups, Planetarium,
1st Department, Star City.
Photograph, 2009.

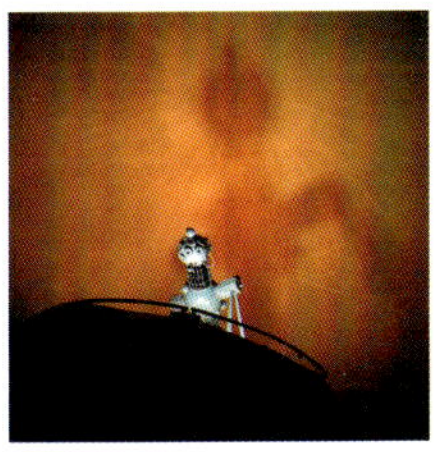

Planetarium,
1st Department, Star City.
Photograph, 2009.

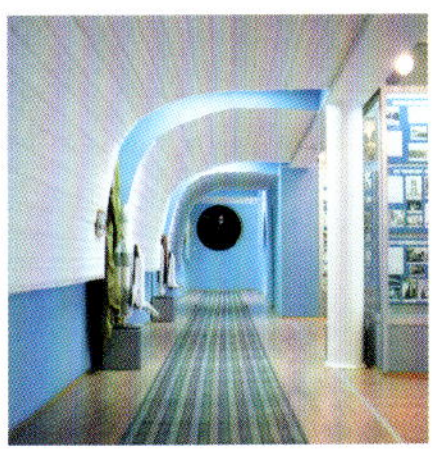

Hallway,
Baikonur city museum,
Baikonur.
Photograph, 2010.

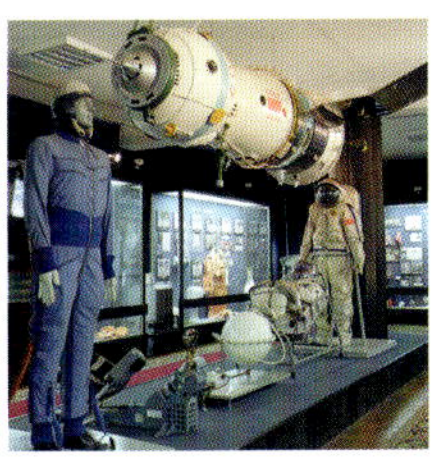

Museum installation,
House of Cosmonauts,
Star City.
Photograph, 2010.

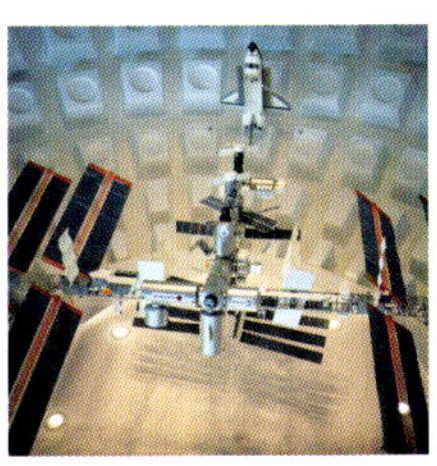

ISS mock-up,
Baikonur city museum,
Baikonur.
Photograph, 2010.

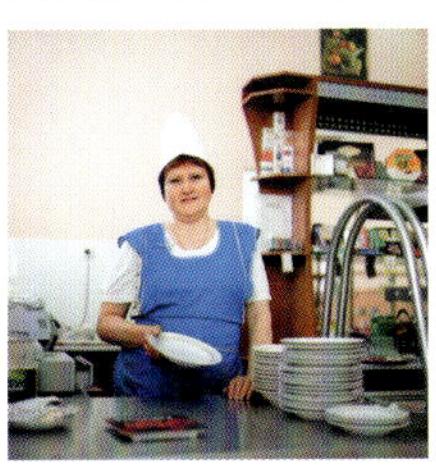

Cook, Fedorova,
2nd Department, Star City.
Photograph, 2009.

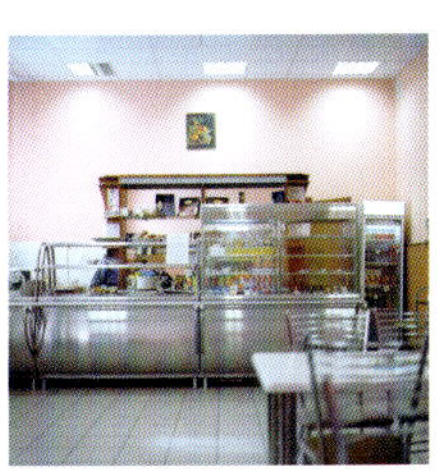

Canteen,
2nd Department, Star City.
Photograph, 2009.

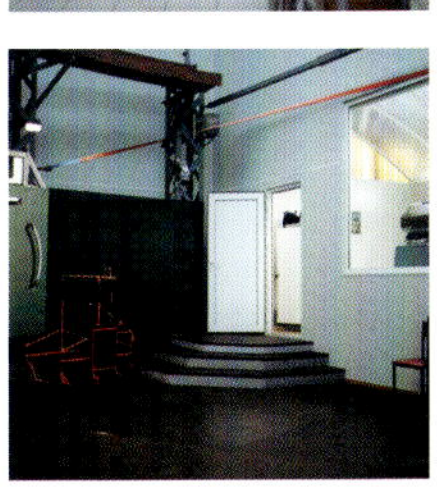

Entrance, control room,
Vykhod specialised
simulator,
2nd Department, Star City.
Photograph, 2010.

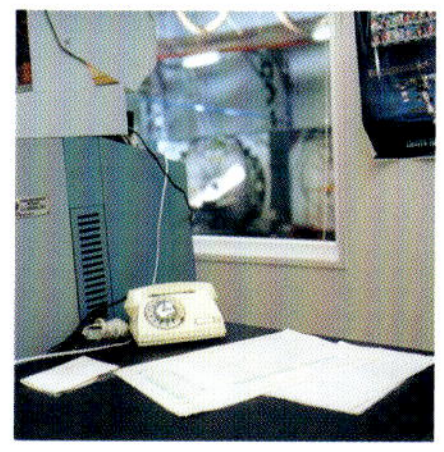

Control room, Vykhod
specialised simulator,
2nd Department, Star City.
Photograph, 2010.

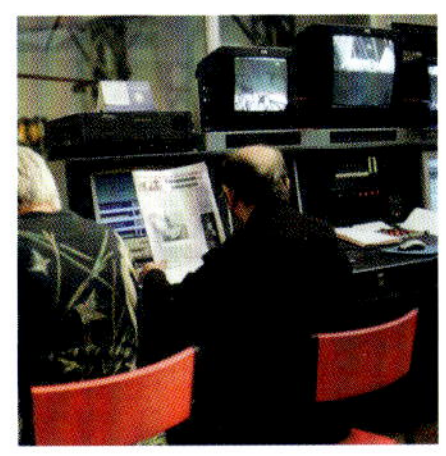

Control room, Vykhod
specialised simulator,
2nd Department, Star City.
Photograph, 2010.

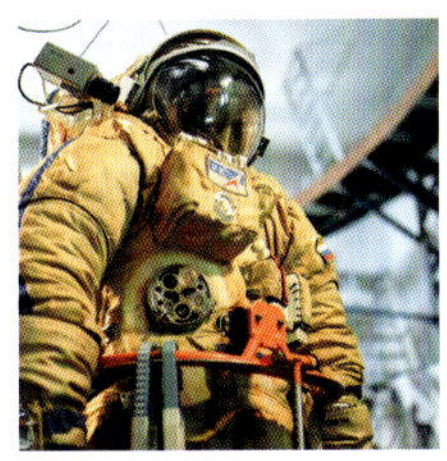

Space suit, Vykhod
specialised simulator,
2nd Department, Star City.
Photograph, 2010.

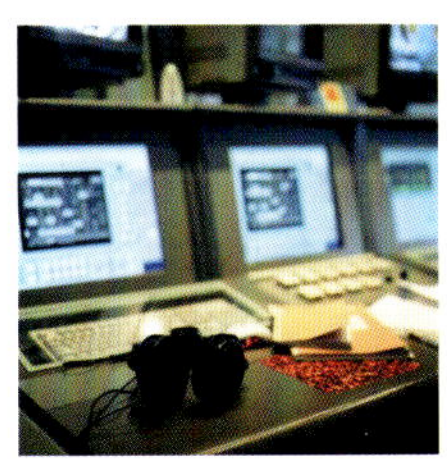

Control panel, Vykhod
specialised simulator,
2nd Department, Star City.
Photograph, 2010.

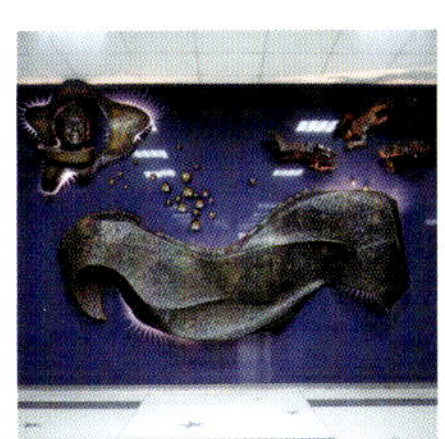

Main entrance hall,
House of Cosmonauts,
Star City.
Photograph, 2010.

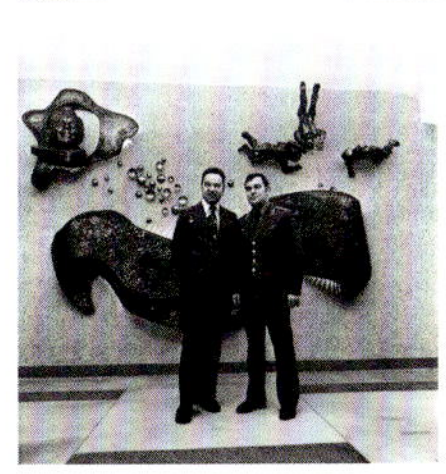

Cosmonauts, Grechko
and Romanenko,
House of Cosmonauts,
Star City.
Archival photograph, 1978.

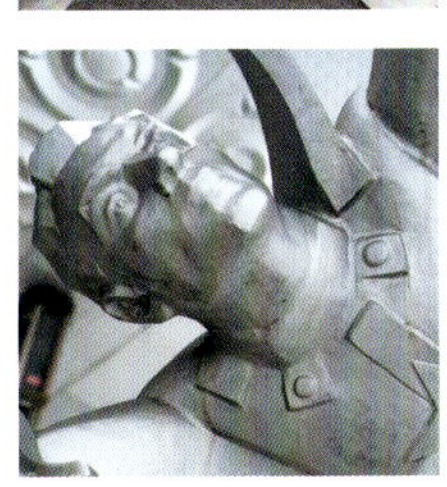

Sculpture of cosmonaut,
House of Cosmonauts,
Star City.
Photograph, 2010.

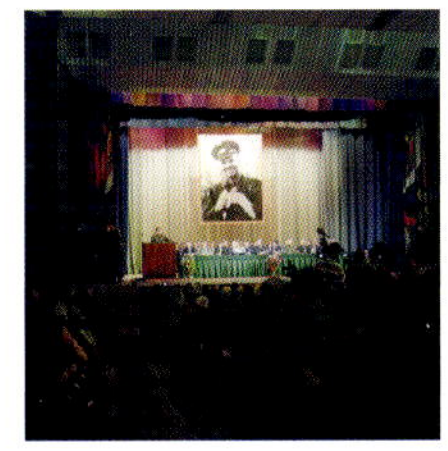

Official ceremony,
Assembly Hall,
House of Cosmonauts,
Star City.
Photograph, 2009.

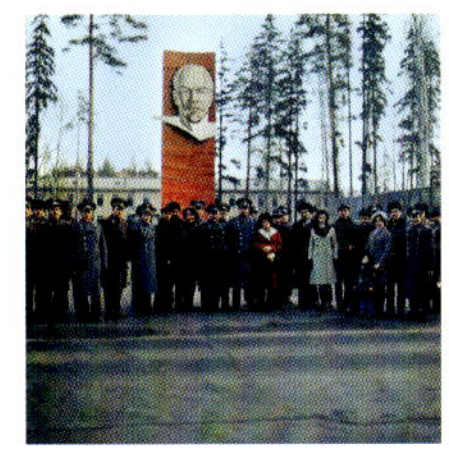

Soyuz–16 space flight crew,
Star City.
Archival photograph, 1974.

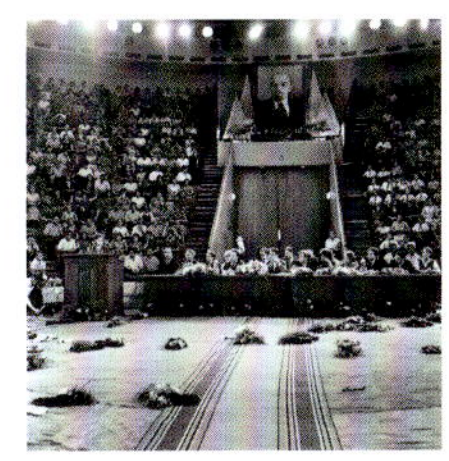

Official ceremony.
Archival photograph, 1963.

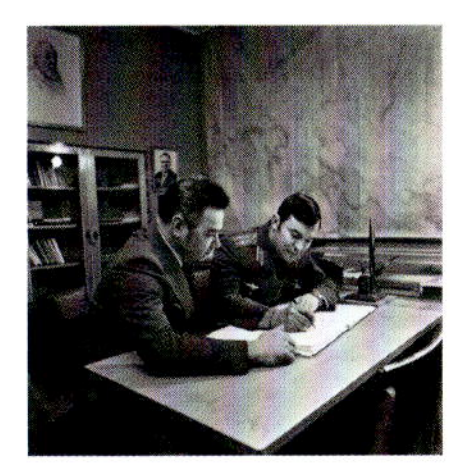

Cosmonauts,
Grechko and Romanenko,
Gagarin's office,
House of Cosmonauts,
Star City.
Archival photograph, 1978.

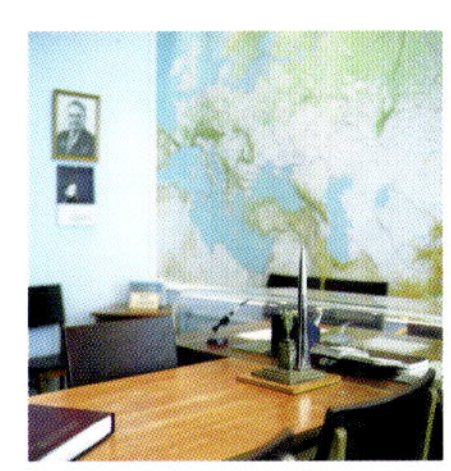

Office of Y A Gagarin,
House of Cosmonauts,
Star City.
Photograph, 2010.

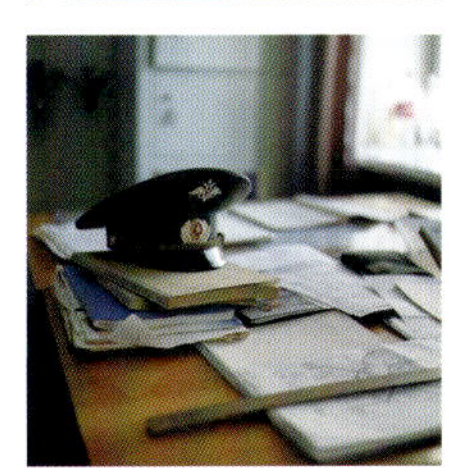

Office table,
3rd Department, Star City.
Photograph, 2010.

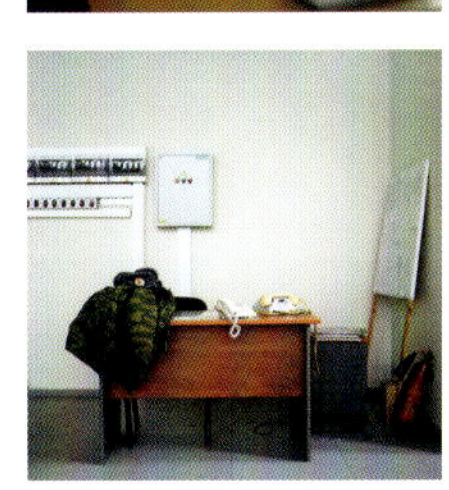

Office table,
2nd Department, Star City.
Photograph, 2010.

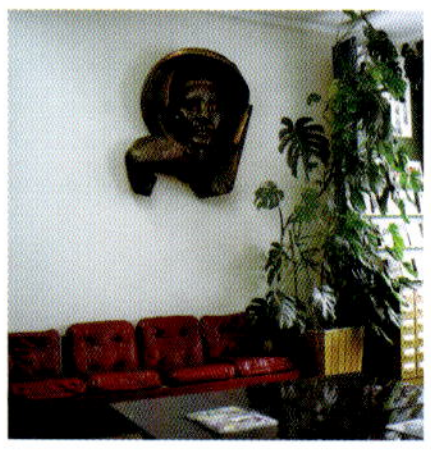

Library, House of Cosmonauts, Star City.
Photograph, 2010.

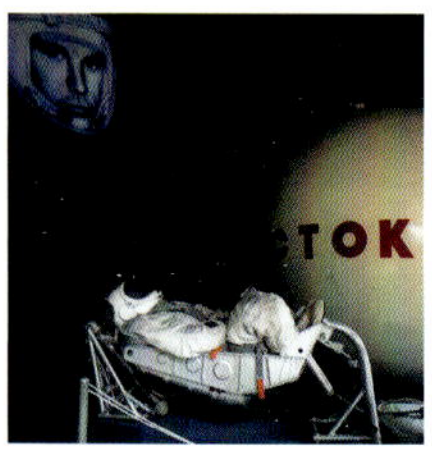

Vostok spacecraft, cosmonaut dummy, installation, Museum of the House of Cosmonauts, Star City.
Photograph, 2010.

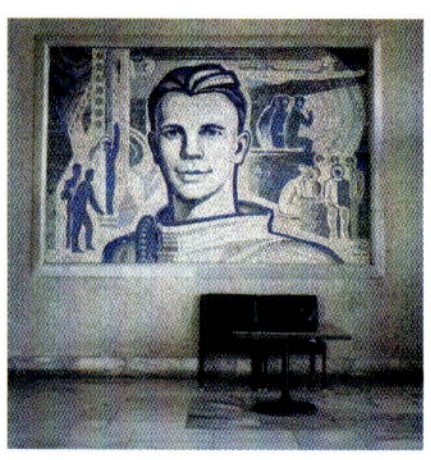

Decorative mosaic, 1st Department, Star City.
Photograph, 2010.

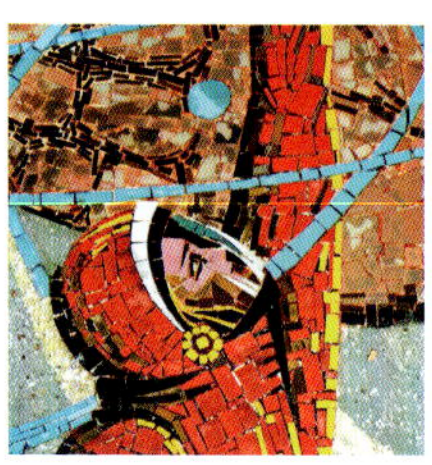

Decorative mosaic, Baikonur city museum, Baikonur.
Photograph, 2010.

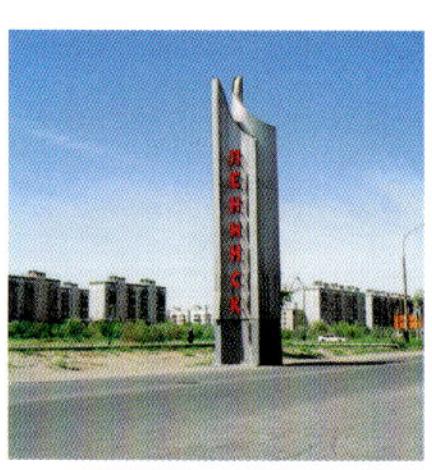

Obelisk Leninsk, Baikonur.
Photograph, 2010.

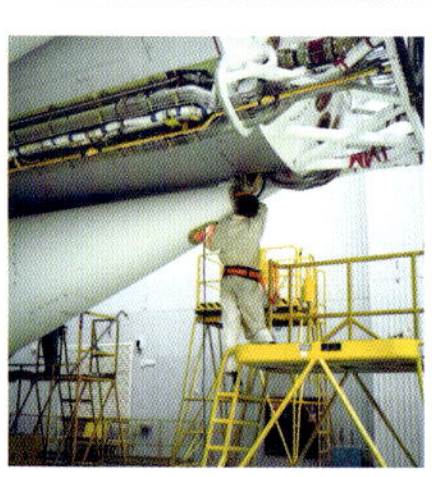

Engineer, Missile Assembly Building, Baikonur.
Photograph, 2010.

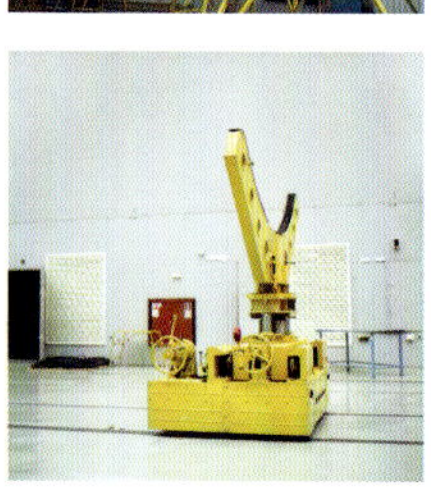

Rocket support prop, Missile Assembly Building, Baikonur.
Photograph, 2010.

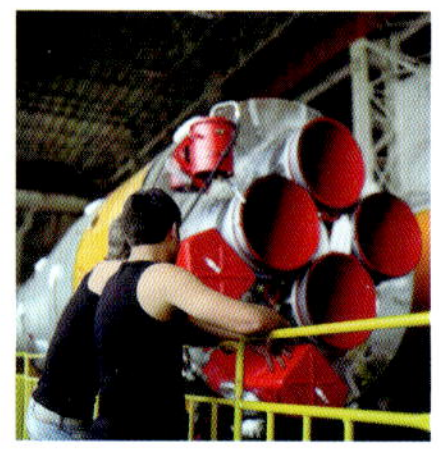

Students at the Missile Assembly Building, Baikonur.
Photograph, 2010.

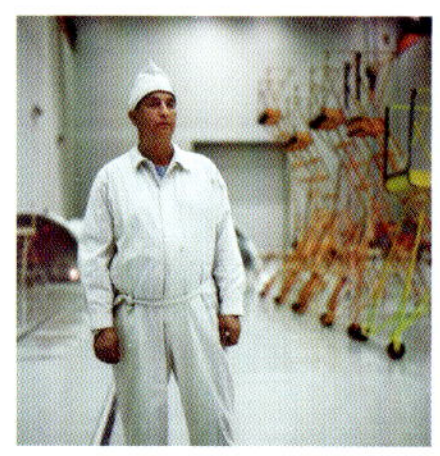

Engineer, Missile Assembly Building, Baikonur.
Photograph, 2010.

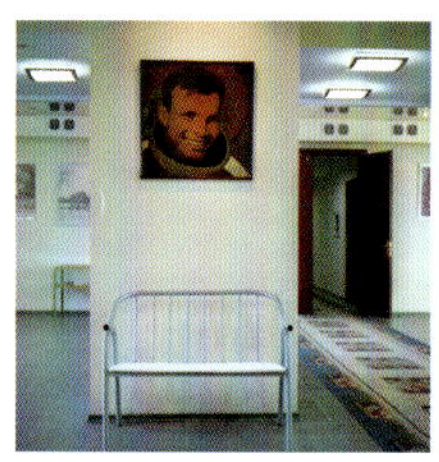

Semi-precious stone portrait of Gagarin, Museum of Gagarin, Baikonur.
Photograph, 2010.

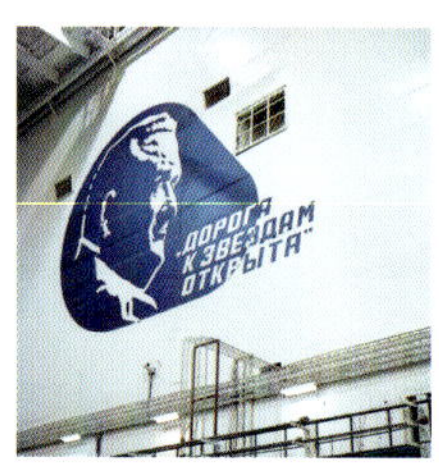

Wall, Missile Assembly Building, Baikonur.
Photograph, 2010.

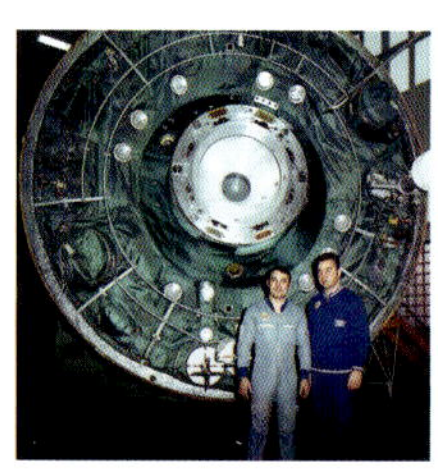

Cosmonauts, Grechko and Romanenko, 2nd Department, Star City.
Archival photograph, 1978.

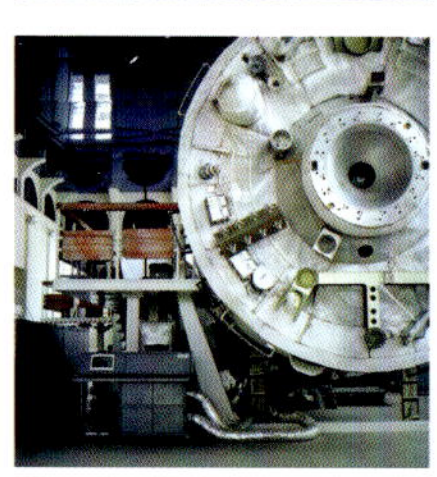

Zvezda service module, 2nd Department, Star City.
Photograph, 2010.

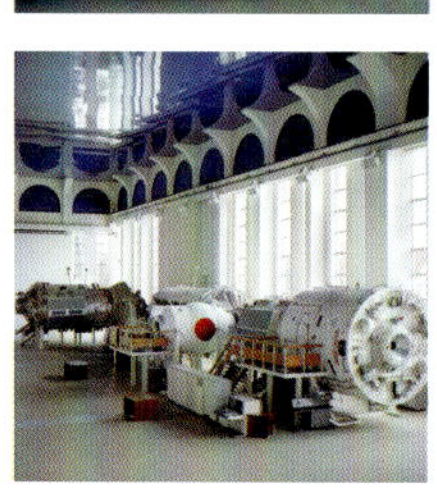

Zvezda service module, 2nd Department, Star City.
Photograph, 2010.

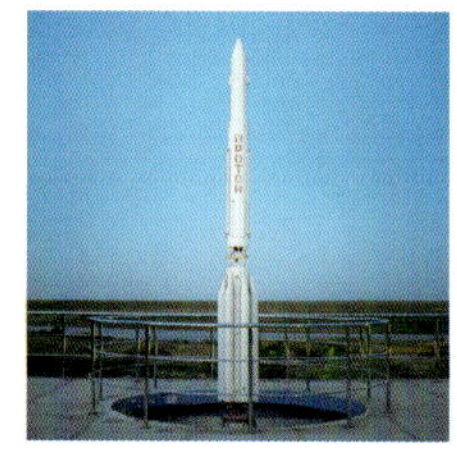

Proton mock-up, Alley of Cosmonauts, Baikonur.
Photograph, 2010.

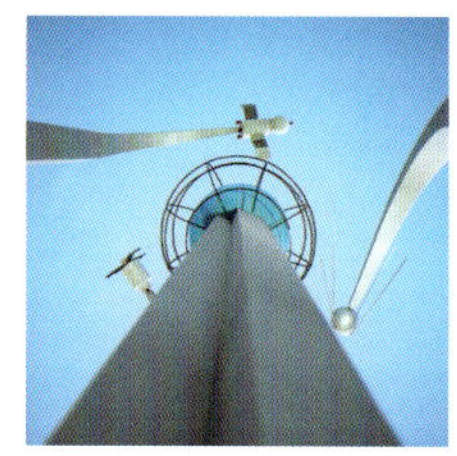

Obelisk, Baikonur.
Photograph, 2010.

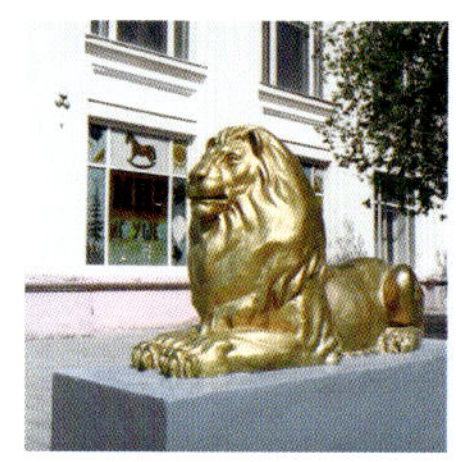

Sculpture and toy shop, Baikonur.
Photograph, 2010.

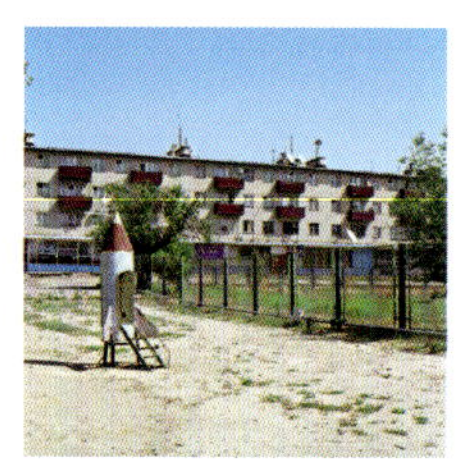

Playground, Baikonur.
Photograph, 2010.

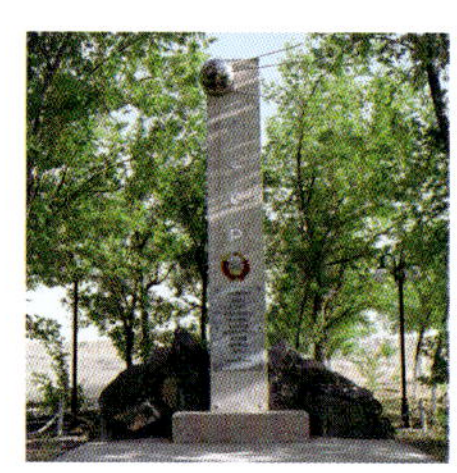

Obelisk, Baikonur.
Photograph, 2010.

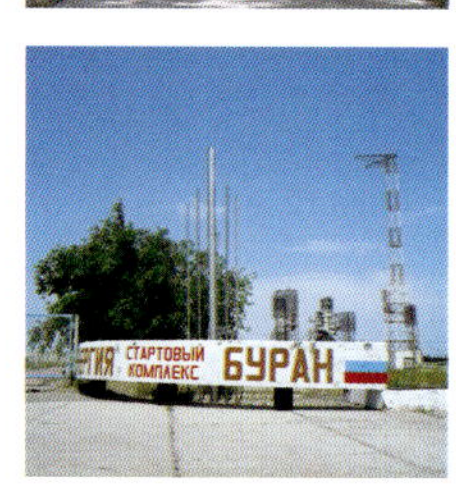

Gates, abandoned Buran launch site, Baikonur.
Photograph, 2010.

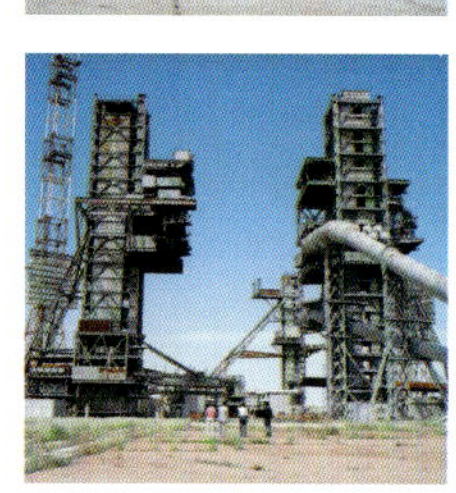

Abandoned Buran launch site, Baikonur.
Photograph, 2010.

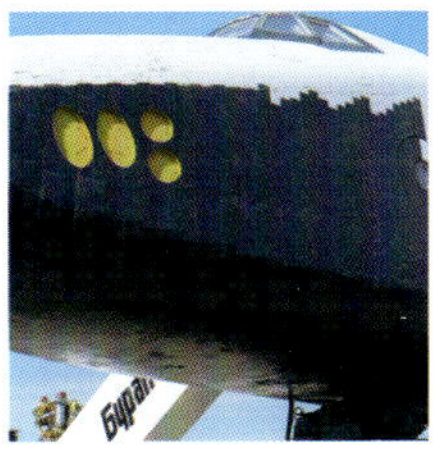

Cabin of Buran spacecraft, installation, Museum of Gagarin, Baikonur.
Photograph, 2010.

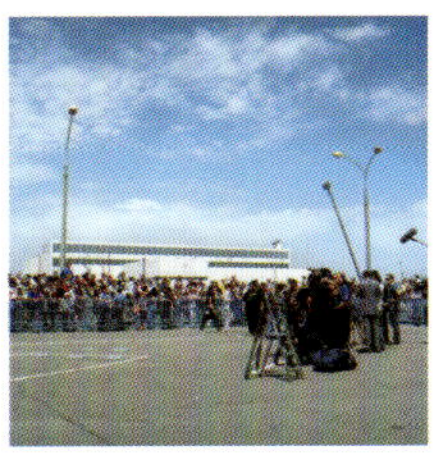

The press waiting for cosmonauts, Baikonur.
Photograph, 2009.

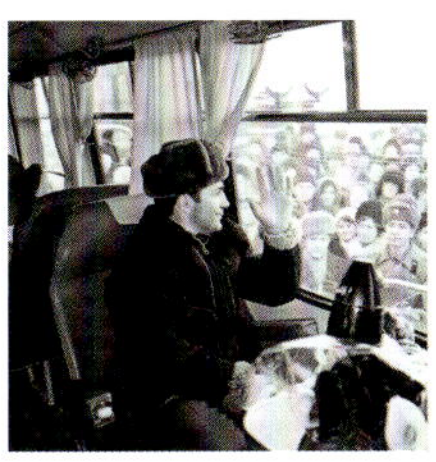

Cosmonaut, Klimuk.
Archival photograph, 1973.

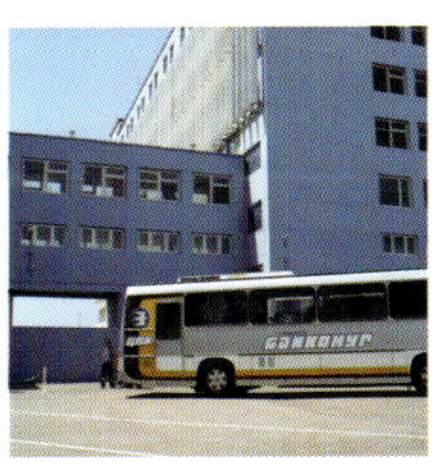

Bus, Baikonur.
Photograph, 2009.

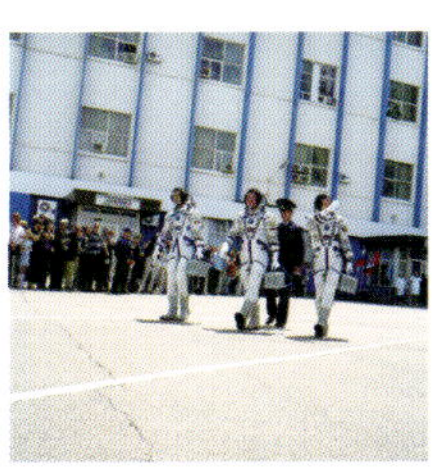

Cosmonauts, Romanenko, Frank De Winne, Robert Thirsk, Baikonur.
Photograph, 2009.

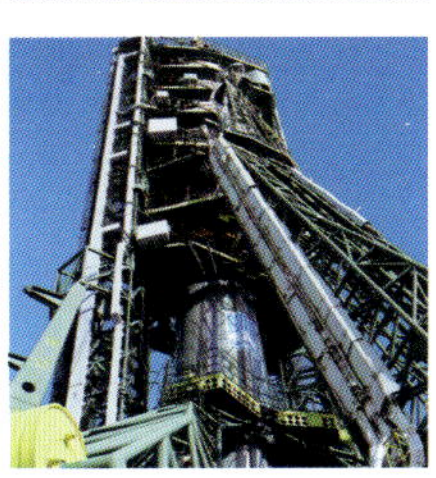

Soyuz spacecraft at launch pad, Baikonur.
Photograph, 2009.

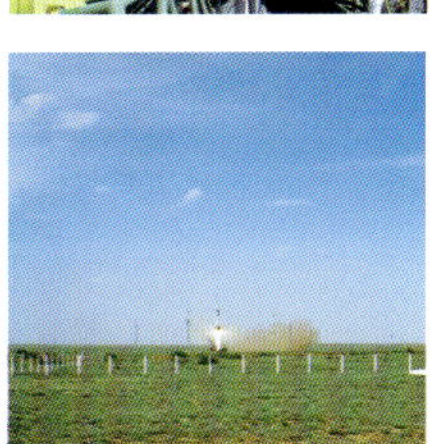

Launching of Soyuz TMA-15 space misssion, Baikonur.
Photograph, 2009.

QUOTATIONS

p.44 Leonov, A A
1974 interview in: *Handshake in Space, Bridge into Space* (1976), Moscow: Izvestia.

p.62 Grechko, G M in:
From Kopernik to Salyut, Interkosmos is Collaboration (1987), Moscow: Mashinostroenie.

p.69 Kiselev, A S
With Gagarin's name: the Yuri Gagarin Russian State Science Research Cosmonauts Training Centre (2005), Moscow: Slovo.

The following cosmonauts feature in the photographs:

V V Aksyonov (b.1935)
A P Aleksandrov (b.1943)
V F Bykovsky (b.1934)
Jean-Loup Marie Chrétien (b.1938)
V A Dzhanibekov (b.1942)
V V Gorbatko (b.1934)
G M Grechko (b.1931)
P I Klimuk (b.1942)
D Y Kondratyev (b.1969)
S K Krikalev (b.1958)
V V Lebedev (b.1942)
A A Leonov (b.1934)
Arnaldo Tamayo Mendez (b.1942)
L I Popov (b.1945)
P R Popovich (1930–2009)
Dumitru Prunariu (b.1952)
Y V Romanenko (b.1944)
V P Savinykh (b.1940)
S Y Savitskaya (b.1948)
A A Serebrov (b.1967)
A N Shkaplerov (b.1972)
A V Solovyov (b.1946)
V V Tereshkova (b.1937)
Robert Thirsk (b.1953, CSA astronaut)
A S Viktorenko (b.1947)
I P Volk (b.1937)
A A Volkov (b.1948)
Frank De Winne (b.1961, ESA astronaut)

I would like to thank Makarov V V, Osipov V B, Perminov A N, Tsbliyev V V, Krikalev S K, Onufrienko Y I, Andrei Ledenev, Alexander Siniagin, Nina Gomiashvili, Alexander Sokolov, Garry Wallis, Dave Hendley, Tim Marshall, Phil Baines, Max Ackermann, Mark Power, David Moore, Paulus Dreibholz, Paul Rennie, Malcolm Parker.

Maria Gruzdeva, January 2011

First published in the United Kingdom in 2011 by

Dewi Lewis Publishing
8 Broomfield Road, Heaton Moor
Stockport SK4 4ND, England
www.dewilewispublishing.com

ISBN: 978-1-907893-05-6

Design and Layout: Dewi Lewis & Maria Gruzdeva
Print: EBS, Verona, Italy

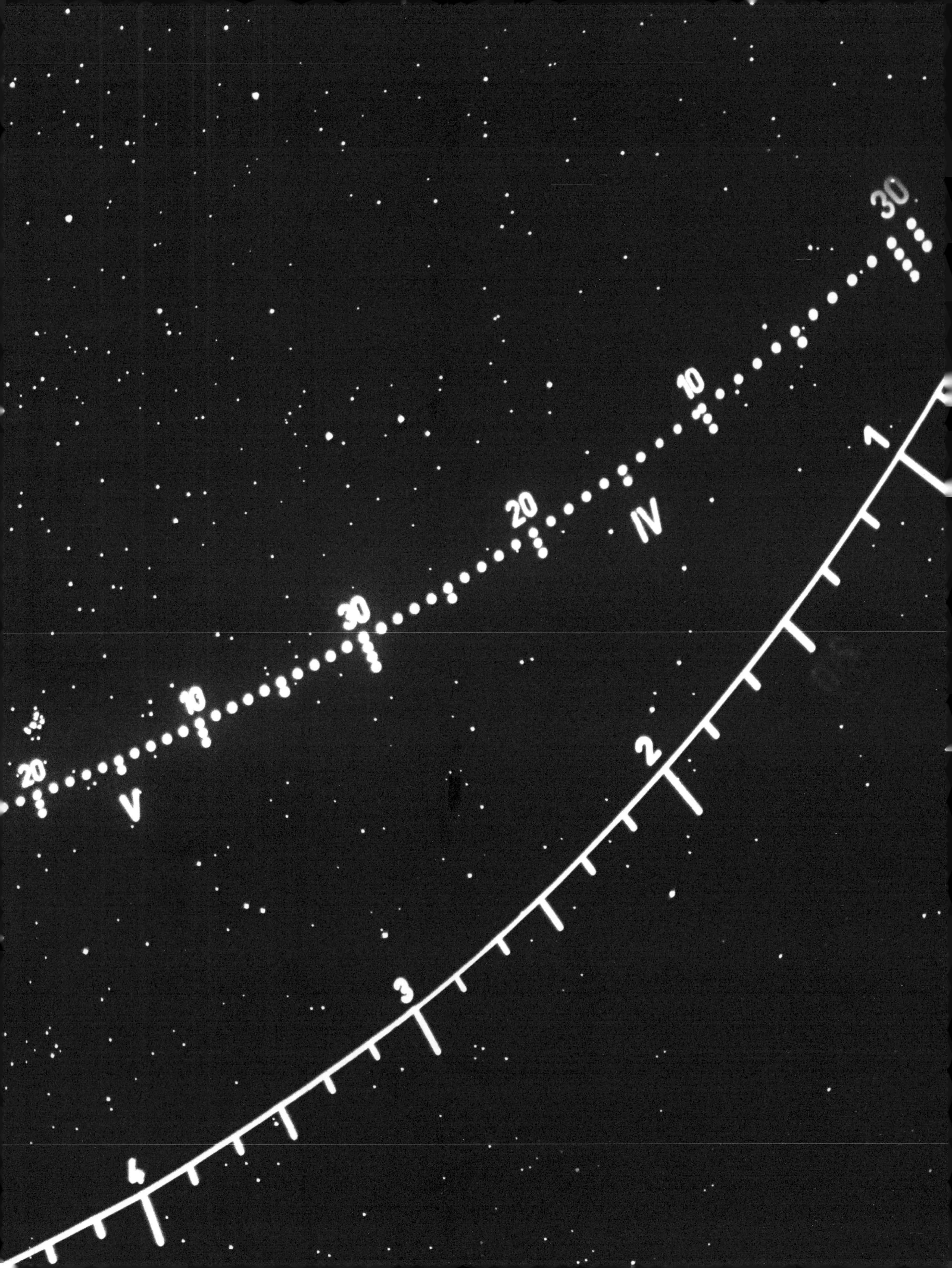

30
10
1
IV
20
30
10
20
V
2
3
4